PACIFIC COAST PELAGIC INVERTEBRATES

A Guide to the Common Gelatinous Animals

by

David Wrobel and Claudia Mills

SEA CHALLENGERS
4 Sommerset Rise, Monterey, CA 93940

MONTEREY BAY AQUARIUM
886 Cannery Row, Monterey, CA 93940

MONTEREY BAY
AQUARIUM®

1998

A SEA CHALLENGERS AND MONTEREY BAY AQUARIUM PUBLICATION

Editor - Dr. Hans Bertsch

Front Cover

Top row:	*Pelagia colorata* - David Wrobel
	Gonionemus vertens - Claudia Mills
Middle row:	*Beroe forskalii* - David Wrobel
	Carinaria cristata - David Wrobel
Bottom row:	*Cliopsis krohni* - David Wrobel
	Pegea socia - David Wrobel

Library of Congress Cataloging-in-Publication Data

Wrobel, David, 1956-

    Pacific coast pelagic invertebrates: a guide to the common gelatinous animals/David Wrobel and Claudia Mills.

    112 p.   cm.

    Includes bibliographical references and index.

    ISBN0-930118-23-5

    1. Marine Invertebrates - Pacific Coast (North America)—Identification.

    I. Mills, Claudia, 1950-   . II. Title.

    QL365.4.P33W76   1998

    592.177'43—dc21

                                              97-42060

                                                 CIP

SEA CHALLENGERS
4 Sommerset Rise, Monterey, CA 93940

MONTEREY BAY AQUARIUM
886 Cannery Row, Monterey, CA 93940

MONTEREY BAY
AQUARIUM®

Printed in Hong Kong through Global Interprint, Petaluma, CA. USA

Typography and prepress production by Diana's Secretarial Service, Danville, CA. USA

ACKNOWLEDGMENTS

The following scientists have been very helpful in reviewing sections that pertain to their own specialties:

Ronald W. Gilmer, Telluride, Colorado
Steven H. D. Haddock, University of California at Santa Barbara
G. Richard Harbison, Woods Hole Oceanographic Institution
Ronald J. Larson, U. S. Fish and Wildlife Service
Lawrence P. Madin, Woods Hole Oceanographic Institution
Philip R. Pugh, Southampton Oceanography Centre, England

Thanks also to Richard Harbison, George Mackie, Jim Childress, Chuck Greene, Edie Widder, and Philippe Laval for inviting C. Mills to participate in their cruises to study open-ocean and deep-water fauna. Hans Bertsch, Judith Connor, George Matsumoto, Dave Powell, Bruce Robison and Steve Webster provided useful comments after reviewing the manuscript. Thanks to Roland Anderson, Marcus Biondi, Peter Fenner, Chuck Galt, Lisa Gershwin, Yayoi Hirano, Douglas Machle, Jun Nishikawa, Pamela Roe, Tom Schroeder and Erik Thuesen for offering their expertise while reading bits and pieces. Thanks to Lisa Bibko, Richard Brodeur, Paul Dayton, Gregg Dietzmann, Per Flood, Ronald Gilmer, Steven Haddock, Richard Harbison, Thomas Heeger, Barry Heller, Ron Larson, Deanna Lickey, Larry Madin, Richard Miller, the Monterey Bay Aquarium and the Monterey Bay Aquarium Research Institute for use of their photographs. C. Mills wishes to acknowledge the support of colleagues and use of the facilities of the Friday Harbor Laboratories over many years. D. Wrobel is very grateful to Leon Garden for being a buddy on many dives photographing jellies and for assistance with numerous collecting trips. Freya Sommer was very helpful in sharing her knowledge of jellies. The Monterey Bay Aquarium provided use of a photomicroscopy system and permitted photography of captive specimens. The Monterey Bay Aquarium Research Institute has been a valuable source of information gleaned from their scientific staff, video library, and public presentations. D. Wrobel also appreciates the support of his wife Debra, who has to put up with his gelatinous and photographic interests.

Many many thanks to the librarians at Hopkins Marine Station, Friday Harbor Laboratories, and University of Washington for all of their assistance in tracking down old literature. We have read as thoroughly as possible in preparing the descriptive and natural history portions of this book. Because of its guidebook nature, the limited space does not allow us to cite the source for every piece of information within the text, which combines our own observations with those taken out of the literature. This book has benefited enormously from the easy communication now offered by use of the Internet. We have had many short discussions by e-mail with other plankton specialists throughout the world over choices of names and various details in this book. This is not to guarantee that everything in the book is correct, but we have tried our hardest to approach that standard. This is an equal-authored book, both of us having contributed equivalently in the final publication.

Although the geographical coverage of this volume is from Baja California to Alaska, we admit that our knowledge is best from British Columbia to southern California and detailed information begins to fall off both to the south and to the north. For the oceanic medusan fauna off Mexico proper (south of Baja California) and central America, those with access to the scientific literature should consult the monograph by Lourdes Segura-Puertas (1984).

PHOTOGRAPHIC CREDITS

All photographs are the copyright of the authors, except for species numbers noted below:
Lisa Bibko: 135
Richard Brodeur: 84
Paul Dayton: 92

TABLE OF CONTENTS

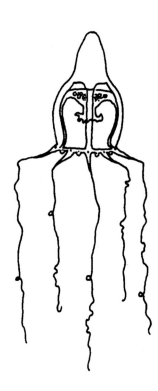

INTRODUCTION

As more people spend time on the ocean, encounters with gelatinous animals have become increasingly frequent. For many, these animals are totally alien and mysterious. To date there has been no comprehensive guide to jellies, comb jellies and other gelatinous zooplankton that inhabit waters off the Pacific coast of the United States and Canada. Only specialists will have had access to the scientific literature on these animals, which is widely scattered and often difficult to find. Non-scientists have had relatively few sources of information.

We have designed this book as a guide to the common gelatinous zooplankton of the west coast that may be encountered by divers, boaters, naturalists, teachers, scientists, students, and others with an interest in marine invertebrates. The gelatinous groups we have included are the cnidarian jellies (hydrozoan, cubozoan and scyphozoan), the comb jellies (or ctenophores), pelagic gastropods, and the pelagic tunicates. (Both Cnidaria and Ctenophora are pronounced in the United States with a silent "c," while most Europeans pronounce the "c.") Although most species in the book are relatively common, we have included some that are rarely seen. Other species are found only in deep water and may be abundant, but rarely observed by anyone other than scientists. We have also included some examples of commensal associations between certain crabs, amphipods and juvenile fishes with their gelatinous hosts.

We could not include all of the gelatinous animals from this region in a book of this size. Species within groups like the siphonophores can be exceedingly difficult to identify, so we have included only a small number of representative species. In some cases we have grouped similar species within a genus under a single entry, with information about others that may be found along the west coast. We have touched only on the most abundant deep-water species; many more inhabit the mysterious dark realm of the midwater. Other gelatinous groups, like the chaetognaths (arrow worms) and pelagic polychaete worms, are not covered in this guide. Despite all the species that we have not included, we believe that most of the time, when you encounter a shallow water gelatinous animal along the west coast, you should be able to identify it using this guide.

The ideal situation for identifying a gelatinous animal is to view it in the natural habitat or in an aquarium. Only then can you observe natural behaviors and see the body in full form. Animals preserved in 5% Formalin can be useful for identification. Preserved specimens may fall apart, however, and can lose coloration. A microscope may be required for small species and to examine the detailed structure of larger animals. A hand-lens or magnifying glass can be useful for examining some specimens. Good photographs may be helpful but it is often difficult to distinguish certain features from a photo if the orientation is not right. When encountering an unknown gelatinous creature, first try to place it into one of the four major groups (cnidarian, comb jelly, gelatinous gastropod, pelagic tunicate) and then go from there.

We have included gelatinous animals that inhabit the cool waters of the Gulf of Alaska to the semi-tropical waters of Mexico's Baja California peninsula. Most are not confined exclusively to this range, as many species are distributed throughout the world's oceans.

Scientific Names and Common Names

All described organisms have a scientific name that is composed of two words, the genus and species. These names are usually derived from Greek or Latin roots and often are chosen to be specifically descriptive of the organism. For instance, the jellyfish genus *Phacellophora* is of Greek origin, meaning "bearer of bundles or clusters," referring to the tentacles. Other genus or species names come from mythological characters or are in honor of scientists who worked in related areas. Scientific names are italicized in print. Many organisms also have common names, such as king salmon, silver salmon, or chum salmon. Since few gelatinous animals are well-known, most do not have common names, but in the few cases where common names are in use, we give both. A full taxonomic list of species included is given near the end of the book. Because classification is an ongoing process, the state of scientific names is in constant flux as more is learned about relationships between organisms. We have given here the most current names for the organisms we present, but a look through the older literature readily reveals just how short-lived many scientific names may be.

What is a Gelatinous Animal?

Many people know jellies and other gelatinous animals only as unidentifiable jelly-like blobs on the beach. Jellies are often treated as a scourge, deadly creatures that should be avoided due to their venomous sting. In reality, gelatinous animals are among the most beautiful and stunning animals of the sea. Although a few types may present some danger to people, most are harmless. None of the jellies that visit nearshore waters of the Pacific coast of North America contain toxins that are capable of killing a person (with the exception of *Physalia* in very rare instances).

Take a jelly out of water and it collapses into a quivering blob. The disparate groups that comprise the gelatinous animals are united by having bodies lacking hard skeletal components. With tissues composed of 95% or more water, their bodies are usually exceedingly delicate and easily damaged, with a gelatin-like consistency. Unlike a fish, which retains its form when out of water, a jelly requires the support provided by the aqueous environment. Gelatinous animals are remarkable in their abilities to swim and capture food without the aid of hard skeletal parts (with the exception of some pelagic molluscs that have jaws and a radula).

The true jellies, or medusae, are members of the Phylum Cnidaria, and consist of hydromedusae, scyphomedusae, and cubomedusae. This diverse and widespread phylum also includes corals and anemones. The scyphozoans are the most familiar of the gelatinous animals, with representatives such as sea nettles and moon jellies. Many are large and conspicuous and typify what comes to mind when people think of jellies. All are equipped with the unique cnidarian stinging structures known as nematocysts (or cnidocysts), which enable them to be effective predators despite their lack of jaws and teeth. Nematocysts contain a variety of toxins and are typically concentrated in the sinuous tentacles and mouth region. Jellies vary in the number and length of tentacles but generally use them to incapacitate zooplankton prey such as copepods, krill, larval fish, invertebrate eggs and larvae, and other gelatinous animals. The rhizostomes, tropical scyphozoans that lack tentacles, generally use their oral arms in a similar manner. Several rhizostome medusae may be found in the warmer waters of the southern range of this book. One species of cubomedusae (the box jellies) occurs locally along the coast of southern California and Baja California, but it does not have a sting lethal to humans like some of its Indo-Pacific relatives. Less familiar to most people are the hydrozoans, which include a bewildering array of smaller and generally inconspicuous species. Our most potent stingers, the siphonophores, are within this group.

Cnidarian jellies typically use their bell for a sort of jet propulsion. By contracting the thin layer of muscle tissue in the bell, the margin is moved inward and water is expelled, propelling the animal forward. Acting as a hydrostatic skeleton, the gelatinous bell springs back to its resting state when the muscle contraction ceases.

The phylum Ctenophora (ctenophores or comb jellies) is only distantly related to cnidarians. Like the true jellyfish, ctenophores are predators on a wide range of zooplankton types. Ctenophores typically have a solid globe-like body, rather than the pulsating bell of cnidarians, and utilize unique rows of cilia organized into a series of plates that beat in an oar-like fashion with highly synchronized waves. They also lack the stinging nematocysts of jellies and use other means for collecting food, including sticky tentacles, mucous-covered oral lobes and voluminous, highly mobile mouths.

Some groups less familiar to most people have also taken up a gelatinous existence. Among the more abundant gelatinous zooplankton are the various types of pelagic gastropods within the phylum Mollusca. Lacking a hard, dense shell and using modifications of the molluscan foot for locomotion, many of these swimming snails barely resemble their benthic relatives. Few people are aware that some of the most common planktonic animals along our west coast are gelatinous snails. Predatory heteropods possess a single swimming fin to scull through the water while seeking prey. Pteropods use a pair of swimming wings, and some are known as sea butterflies due to the manner in which they flap them. One group of pteropods uses various styles of mucous nets to graze on phytoplankton and other small particles in quiet oceanic water. Another group is predatory, using complex mouth structures to capture gelatinous prey.

The tunicates are known to most people only by their sessile benthic form, the sea squirts. Tunicates are a subphylum within the phylum Chordata, to which humans and all other vertebrates also belong. Two groups of less familiar tunicates have taken to the open sea for their complete life cycle. The thaliaceans include salps, doliolids and pyrosomes. Salps are remarkable grazers that are among the fastest growing of all organisms and are capable of forming incredibly dense populations by a combination of rapid sexual and asexual reproduction. Salps and doliolids employ sets of ring-like muscle bands in their cylindrical bodies. By contracting these muscles while regulating the opening and closing of a pair of body openings, the pelagic sea squirts swim quite efficiently by jet propulsion. Another group of tunicates, the tadpole-like appendicularians (or larvaceans), are generally inconspicuous but ecologically important grazers that use a mucous filter to collect microscopic planktonic organisms. Swimming and water filtration are accomplished by rhythmic beating of the tail.

Properties of Gelatinous Animals

At first glance it might seem to be a rather foolish strategy for a group of animals to have a gelatinous lifestyle. Being somewhat slow swimmers with easily damaged tissue and no hard skeletal parts might appear to be a clear path to extinction. The abundance and wide diversity of gelatinous animals is testimony, however, to the numerous advantages that this lifestyle presents. When thinking about jellies we need to suspend our bias toward hard skeletons with thick muscles and other relatively dense tissues.

With bodies that lack bone and thick muscles, gelatinous animals have come up with a solution to the problem of maintaining buoyancy in the water column. Rather than constructing air bladders, harboring lipid-dense bodies or using energetically expensive swimming, jellies in essence become one with their environment since they are mostly water. While some gelatinous animals (such as certain pteropods that retain a remnant of the shell) must swim to some extent to compensate for the extra weight, this requirement is far less than would be required for a comparably-sized denser animal. Buoyancy can also be enhanced by the exclusion of certain ions (primarily sulfate) or accumulation of others (ammonium) in body tissues. Lipids, which are less dense than water and stored by zooplankton such as copepods for buoyancy, are not typically used by gelatinous animals.

Gelatinous zooplankton have a far higher water content on a percentage of mass basis than other marine creatures such as fish or crustaceans. With their high water content and relatively low carbon density, there is a greater potential during growth for rapid increase in size on a volume basis. This helps render them a relatively poor source of food for predators lurking about. The low carbon density enables a low metabolic rate to be maintained. This can be an advantage in that food requirements are lower. As a consequence, many gelatinous animals can withstand long periods without food, sometimes shrinking in size during these periods. When presented with an abundance of food, however, many jellies are opportunistic and can respond with rapid increases in size or population density. Certain species can also regenerate parts of their bodies that have been damaged or bitten off by a predator.

A large surface area relative to carbon content also results in an increased area for capture of prey and gas exchange. Certain medusae and siphonophores act like living drift nets, exposing a large tentacle network to passing zooplankton prey. Larvaceans and thecosome pteropods utilize large, delicate mucous webs to gather tiny food particles. With their thin tissues and high surface area exposed to the environment, cnidarian jellies and ctenophores have no need for specialized gas exchange structures. Pelagic gastropods and tunicates may have reduced gills but can also exchange dissolved gases over much of the body surface. Associated with the gelatinous lifestyle are severe osmotic problems in dilute ionic environments. Fortunately for them, with the exception of a few species adapted to fresh and brackish water, most will never encounter such situations.

Another property of gelatinous tissue is that it is often transparent to light. Transparency is useful for avoidance of visual predators in sunlit open water habitats that lack shelter. This is particularly important for delicate gelatinous creatures since the absence of hard body plates and the relatively poor swimming ability renders most quite susceptible to predation. While medusae, comb jellies and pelagic molluscs may have limited escape responses, they are still slow compared to fishes and turtles that are seeking a meal. Only the cnidarian jellies with their stinging nematocysts can mount any serious defense against predators. With their capable defenses, many scyphomedusae such as purple-stripe jellies and sea nettles have abandoned the transparency strategy.

Gelatinous cnidarians, ctenophores and thaliaceans have relatively low carbon to dry weight ratios (less than 15%). Some of the pelagic molluscs are more semi-gelatinous, with values ranging between 20 to 30%. For the purposes of this book, we will consider all of these species to be gelatinous.

Role of Gelatinous Zooplankton in Marine Ecosystems

Gelatinous zooplankton are ubiquitous in marine environments, inhabiting all the oceans and ranging from coastal nearshore habitats, to the open ocean and down into the deep sea. Despite their delicate consistencies, many play crucial roles in various marine ecosystems, serving as predators, grazers and sources of food for other animals. With rapid generation times and high growth rates, many species can rapidly take advantage of local concentrations of zooplankton or phytoplankton. In a matter of days, profound ecosystem shifts can occur within an area as a result of the predatory or grazing activities of gelatinous animals.

Long tentacles, large oral lobes or high surface area mucous webs are utilized by many species to passively capture other kinds of plankton. These modes of food collection can be very effective in causing substantial reductions in populations of zooplankton prey such as copepods and larval fish. The drift net method is also employed by deep-water siphonophores, which are among the premier predators of the midwater zone. Although many medusae, siphonophores and ctenophores are relatively weak swimmers, relying primarily on currents to transport them within food-laden areas, they are often capable of controlled movements that enhance prospects of entering and remaining in water masses with high zooplankton densities. Small-scale, highly stereotyped swimming patterns can be used to deploy tentacles in set configurations, while also maintaining position in the water column. Other medusae and ctenophores, and thecosome pteropods, are neutrally buoyant and drift passively.

Active predators include certain gastropods (heteropods and gymnosome pteropods), many siphonophores, and beroid ctenophores. Both groups of molluscs use well-developed eyes to seek out their prey, which are captured with complex mouth structures. Some siphonophores swim actively, setting their tentacles frequently; other species are quite passive. The beroids swim actively while cruising for prey, but seem to contact their potential meals blindly before responding–whether or not beroids use other senses to detect nearby prey is not known. These active predators generally are less abundant and have a more limited ability to cause significant ecosystem shifts compared to the more passive species.

Salps include some of the prime grazers within oceanic waters. With their relatively large sizes, rapid rates of pumping water through the body, and high growth and reproductive rates, they can quickly assume numerical dominance within localized areas. Due to their ability to clear the water column of phytoplankton, salps can probably severely reduce food available for other grazers such as copepods and krill. Their fecal pellets are also an important source of food to deeper water communities.

Gelatinous animals have generally been overlooked as major players within the marine realm. During the past decade this notion has shifted to a realization that learning about the ecology of these creatures is crucial to any understanding of the workings of all marine ecosystems.

Habitats and Distribution

Although there are a few freshwater hydromedusae, the vast majority of gelatinous animals are confined to marine habitats. Some can be found in estuaries which may be brackish and vary considerably in salinity and temperature. Species found within such habitats usually have relatively wide tolerances for environmental variation. Most of the species featured in this book are adapted for more open environments, either near the coast or farther out in the open ocean. They generally do not tolerate major shifts in salinity from that of normal seawater (33 to 34 parts per thousand), and may also have limited ranges of temperature tolerance.

If you visit the ocean and look long enough, chances are good that you will see some type of gelatinous animal. Some may be jellies that favor nearshore coastal environments. Others may be more oceanic species that make only occasional (and usually lethal) sojourns to shallow waters. You may also occasionally encounter a species that normally is found much farther to the south in sub-tropical waters. Because these animals reside in an environment that is constantly on the move, it is difficult to pinpoint the distribution of oceanic species with the accuracy accorded to benthic animals. Wind, currents, upwelling fronts and other factors serve to move large masses of water and associated zooplankton. The ranges presented in this book are therefore only general guides to where species may be found. You may go long periods without seeing a particular species, but as a consequence of unusual oceanic conditions or more predictable seasonal changes, a previously rare species may suddenly be found in abundance for days or weeks before fading again into obscurity. In general, temperature tolerances place rough limits on latitudinal and depth ranges.

Close to shore, most of the species found will be those adapted for more highly productive coastal habitats. Most are highly seasonal in appearance, and are typically accustomed to periods of high phytoplankton and zooplankton densities. Some, particularly a number of bottom-associated hydromedusae, are confined to calm waters of protected bays and inlets. Other species may at times be found in these relatively placid habitats but will also venture into offshore coastal areas. Within the fjords and passages of Washington, British Columbia and Alaska, and other protected bays on the outer coast are found numerous species of medusae, siphonophores and ctenophores. These and other species may also travel within the less protected nearshore waters that form most of the coastline from Baja California to Washington. Here they must contend with waves, surge and occasional encounters with the bottom. Kelp forests, with their hordes of hungry animals and dense, entangling thickets of kelp, also present hazards for wandering jellies.

Miles out to sea, beyond coastal influence, is the realm of the truly open ocean creatures. Here the water is less productive, with sparse phytoplankton, and takes on a remarkable, attractive deep blue appearance. Epipelagic oceanic species are those that are generally confined from surface waters down to 100 to 200 meters (the approximate limit of the wind-mixed upper layers and of the photic zone beyond which sunlight penetration is not sufficient to support photosynthesis). These blue-water jellies do not have to contend with hard boundaries such as the bottom or kelp forests. The oceanic environment is incredibly vast, encompassing about 97% of the world's living space. Many gelatinous animals are adapted to this poorly understood habitat. Various salps, pteropods, heteropods, ctenophores and medusae that favor oceanic waters are occasionally carried by currents to nearshore environments. It is at such times that you can observe these gelatinous wanderers, but many are easily damaged by the energetic nearshore environment. More typically you may need to travel at least several miles out to sea to view them.

Midwater habitats below the photic zone harbor numerous gelatinous creatures. Whereas abundance drops with lower food supply, species diversity is actually higher in the midwater zone. Deeper water species are adapted to cold water, generally less than 42°F (5°C). Some migrate into shallow water to feed at night and can adjust temporarily to warmer temperatures. These are the midwater species that are most likely to make their way to shallow nearshore habitats where they can be found by casual observers. Other species remain confined to deeper water and cannot adjust to warmer temperatures. They may also have physiological constraints, such as enzyme systems adapted to high pressure environments and low tolerance for the high oxygen levels near the surface. Many are incredibly delicate since they do not have to contend with the surge and wave effects of shallow water. A number of deep-water medusae and ctenophores have brilliant red to dark red pigmentation, presumably to mask bioluminescent prey that are ingested. In general, only scientists with access to submersible vehicles are able to view these types of gelatinous animals in their natural condition since they rarely, if ever, come near the surface and many are torn beyond recognition by trawl nets.

Along the west coast of North America, within a few miles of shore, the water is generally cool throughout the year, at least as far south as Point Conception in California. The primary influence is the California Current, part of the large-scale clockwise motion of water in the north Pacific. This current brings cool water from the north and keeps coastal surface waters typically below 55°F (13°C) throughout most of the range of this book. Southern California and Baja California coastal waters are less under the influence of the California Current and tend to warm up a bit during the summer and fall. From spring through late summer, prevailing northwesterly winds tend to drive the California Current. Fall brings a slackening of these winds, and more likely intrusions of warmer oceanic water. Winter months are marked by a weakening or reversal of northwesterly winds and a diminution of the current flow. A northward-directed countercurrent may then form, extending from Baja California to beyond Point Conception. This warming influence is produced by the Davidson Current, which typically runs below 200 meters but may reach the surface in the winter. Depending on the latitude, upwelling of deeper nutrient-laden water occurs in spring and summer months. Combined with the phosphate-rich California Current, this causes the phytoplankton blooms that convert coastal waters to various shades of green and brown.

Gelatinous animals are often at the mercy of large-scale motion brought about by currents, which can sweep jellies over long distances and cause dramatic shifts in offshore conditions within days. In areas like Monterey Bay, Puget Sound and the San Juan Islands of Washington, and the straits between the eastern coast of Vancouver Island and the mainland, coastal geography and tidal conditions can significantly influence current motion, so that the prevailing motion in some local areas may not always be from north to south.

Species of gelatinous animals vary in the range of temperatures each can tolerate. In general, coastal cold-adapted species will be confined to the California Current in the range from southeast Alaska and British Columbia to central California. Beyond the boundaries of the current farther out to sea, the oceanic water warms considerably (at least off California but less so to the north). Species that favor warmer water will thus tend to prevail in the nutrient poor, oceanic central Pacific gyre beyond coastal waters. Coastal waters of southern California and Baja California also harbor more warm-tolerant species. These are not hard and fast rules, however. Particularly during the winter, but certainly possible throughout the year, warm-water species can make occasional appearances in coastal areas north of Point Conception. Intrusions of warm oceanic water offer opportunities to look for gelatinous animals that would normally not be present. This type of event is more likely during periods of so-called "El Niño" (El Niño–Southern Oscillation, or ENSO), when the eastern Pacific may be warmer than normal for many months. Patterns of mixing of oceanic and California Current coastal water are very complex and cannot be predicted with much precision.

For the reasons described above, land-bound observers will notice that many gelatinous organisms have sporadic periods of abundance and absence. Near-surface oceanic forms will be most abundant following periods of on-shore winds. Deep water animals may appear in surface waters during periods of upwelling. And many coastal jellyfish have a complex life cycle, with benthic forms releasing young medusae only in the spring or summer when food is abundant. Most of these medusae have lifespans on the order of weeks to months, so are seen primarily from spring through autumn.

Distinguishing the Major Groups

At first glance, there is a bewildering array of gelatinous creatures, and it may be difficult to figure out what you are looking at. If, however, you pay attention to certain key features of a gelatinous animal, you should be able to determine its group affinities. The following are some of the diagnostic characteristics you can use to narrow down your determination.

- Single radially symmetrical swimming bell with marginal tentacles and flap-like velum around margin. Tentacles with nematocysts. Usually small and relatively transparent with little pigmentation and mesoglea that lacks cells. Gastric cavity a simple sac without separate pouches. Predators on various types of zooplankton.
  Phylum Cnidaria/Class Hydrozoa – Subclass Hydromedusae

- Polymorphic individuals in the form of a chain, some species exceeding 30 meters in length, but usually only a few centimeters long. Consists of one or more swimming bells (one group, the cystonects lack these structures), feeding individuals (gastrozooids), protective individuals (dactylozooids), reproductive individuals (gonozooids) and leaf-like, usually transparent bracts. Some with an apical gas-filled float. Long tentacles, which may be difficult to see, often have a potent sting. Predators on zooplankton and fish larvae.
  Class Hydrozoa – Subclass Siphonophorae

- Single radially symmetrical swimming bell with marginal tentacles and no velum, but usually with scallops at margin. Often large, conspicuous, and with pigmentation. Thick mesoglea that may contain cells. May have long, frilly oral arms that lead to mouth. Gastric cavity with nematocyst-bearing filaments and partitioned into pouches. Tentacles with nematocysts, some species have an uncomfortable sting. Predators on various types of zooplankton, often including other medusae.
  Class Scyphozoa (Scyphomedusae)

- Body with 8 rows of combs (ciliary plates), often seen as shimmering waves of color. May have tentacles, but these are not arranged around a bell margin. If present, tentacles nearly always with sticky colloblasts rather than nematocysts. Usually transparent with little, if any, pigmentation (except for some of the deep-water forms). Predators on various types of zooplankton.
  Phylum Ctenophora (Comb Jellies)

- Elongate body with single ventrally placed swimming fin, which is held upward. Sculling motion of the fin moves the animal forward. Well developed pair of eyes on a snout-like head. May have a coiled shell, or the remnant of a shell; others lack a shell completely. Active predators on salps, doliolids, chaetognaths and other gelatinous zooplankton. **Phylum Mollusca/Class Gastropoda – Superfamily Heteropoda**

- Body with large pair of lateral plate-like extensions of the foot. Flapping action of these plates as wings propels the animal. Uses mucous web to passively gather planktonic food particles while drifting motionless. Some with a calcareous shell, others with a soft pseudoconch. **Class Gastropoda – Order Thecosomata (Shelled Pteropods)**

- Body with a relatively small pair of lateral muscular wings. Swims relatively rapidly using a quick beating motion of the wings. Body lacks any kind of shell. Distinct head with two pairs of antennae that may be retracted. Head terminates with a buccal (oral) apparatus housing a radula, specialized hook sacs and a jaw. Active predators on thecosome pteropods. **Class Gastropoda – Order Gymnosomata (Shell-less Pteropods)**

- Body with incomplete circular bands of muscles (various numbers depending on the species), and an anterior and a posterior opening. Muscular pulsing of body wall pumps water through an internal mucous net that gathers tiny planktonic food. Locomotion by jet propulsion. May occur as single individuals or chains of asexually produced individuals (alternate sexual and asexual generations). **Phylum Chordata/Class Thaliacea – Order Salpida**

- Relatively small, transparent body with complete bands of circumferential muscles (8 or 9), and anterior and posterior openings. Feeds on planktonic particles using currents created by cilia rather than pulsing of body. Hangs motionless until disturbed, and then exhibits a characteristic jumpy motion. Complex alternation of asexual and sexual generations. **Class Thaliacea – Order Doliolida**

- Body in the form of a "tadpole" with tail containing a notochord. Forms mucous house which usually surrounds the animal and collects microscopic planktonic particles for consumption. The inconspicuous animal can be seen inside or beside the much larger mucous house as it beats its tail to create feeding currents. **Class Appendicularia/Larvacea**

Precautions to Take With Jellies

Despite the prevalent notion that jellyfish are deadly creatures to be avoided at all cost, most gelatinous creatures present little danger to people. None of the jellies that frequent our coast are deadly in the manner of some of the tropical box jellies of the Indo-Pacific. Ctenophores, pelagic molluscs and pelagic tunicates lack the endogenous nematocysts of the cnidarians and thus are no threat to people. However, at least one species of pteropod (*Creseis*) can be a skin irritant to divers and swimmers when abundant.

Most cnidarian jellies along the west coast are not dangerous to people. Although most species use nematocysts for prey capture, they have generally evolved for capturing small prey and are not capable of penetrating human skin, or at least we do not respond to the toxin if they do penetrate. A nematocyst is a capsule composed of a keratin-like material, with a little door-like structure that opens and everts a long microscopic tube. There are numerous types, which vary in the structure of the tube and possession of spines at the base. Viewed with a microscope, nematocysts are quite imposing structures. Most are found in groups on the tentacles, with smaller numbers surrounding the oral area and in gastric filaments. A tiny barb known as the cnidocil, which can be affected by mechanical or chemical stimuli, triggers the nematocyst to fire. The capsule contains toxin that is then injected through the tube into the epidermis of the prey. Multiplied by thousands, the action of combined nematocysts can be formidable. Some of the organisms within the range of this book produce toxins that can stimulate pain receptors and also cause the release of histamines and other agents of swelling, burning and redness. More severe toxins can have neurological, cardiac and respiratory effects.

It is always prudent to be careful when handling any cnidarian jelly, even if its sting is mild. If you touch a jelly, you may not feel any pain on your hands, but contact with your eyes, wrist or other sensitive areas could be a different story. It is also possible to develop a sensitivity to the toxin, with severe allergic reactions following further contact. Situations that previously provoked no response may then erupt into a more serious condition. For these reasons, it is best to avoid contact with jellies by not picking them up or by wearing some type of hand protection. Jellies sitting on the beach should not be touched since nematocyst activity can persist even after the animal is dead.

Some west coast cnidarian jellies do indeed pack a punch. This is particularly true of many siphonophores, including the infrequent visitor *Physalia* (the Portuguese man-of-war). Often the nematocyst-laden tentacles are difficult to see on the beach or in the water. Their presence is then revealed by intense pain and a series of red welts, with possible headache, shock, faintness, cramps, nausea, vomiting, chills and fever. More than likely such an encounter will ruin your day. Swimmers off the Pacific coasts of southern California and Baja California, and in the Sea of Cortez are most likely to encounter this and other truly painful types. Warmer water also encourages less body protection, resulting in more chances of contact. Fewer people swim in the waters north of southern California, and divers wear full wetsuits or drysuits, so painful encounters with jellies are less likely. Unprotected areas on the face, such as around the mouth, are still fair game for nematocyst attack. Swarms of moderate stingers such as *Chrysaora*, *Pelagia*, *Stomolophus* and *Cyanea* thus can be a nuisance for divers or snorkelers along the entire coast.

Swimming within a swarm of sea nettles or other types of jellies can be a mind-boggling experience. In a dense aggregation, however, it can be difficult to avoid contact with tentacles. A wetsuit will easily prevent nematocyst penetration for areas covered. If you do get stinging tentacles on bare skin, the best thing is to rinse the area with seawater. All traces of tentacles should be removed since nematocyst action will continue long after the initial contact. Freshwater and rubbing alcohol are best avoided for rinsing since they tend to trigger the nematocysts. A popular folk remedy is the use of human urine, but this too has the undesirable effect of triggering the nematocysts. Care should also be taken when handling your wetsuit or scuba equipment after contact, since pieces of tentacles may remain on these items. Even an anchor line may be contaminated with stinging tentacles if a dense swarm is present. Workers on fishing boats in northern waters are very familiar with the tentacles of scyphomedusae being whipped around the deck as the nets are reeled in.

Contact with moderate stingers may result in a burn-like rash on sensitive areas. Redness and some swelling may occur, along with mild to moderate pain, throbbing, prickly sensations and itching. You should resist the temptation to rub the area since this will only cause more nematocysts to discharge. Without treatment, symptoms will usually disappear within a few hours. Pain can often be reduced with cold therapy by applying ice packs or ice wrapped in a towel to the affected area for at least 15 minutes. Large areas may be gently rubbed with a piece of ice, in a plastic bag, held with a glove or cloth while keeping the rest of the body warm with blankets or clothing.

More potent stingers such as *Physalia* have longer term effects. Moderate or severe pain may last for several or more hours, and welts or lesions may take days to disappear. Along with a thorough seawater rinse, stings can be treated with a vinegar or sodium bicarbonate rinse to disable the nematocysts, followed by application of topical analgesics and steroids. Meat tenderizer is often suggested for pain relief, but probably has limited benefit. A more severe reaction, with associated respiratory and cardiac distress, will require prompt emergency treatment by a physician. In such cases intravenous analgesics, antihistamines and hydrocortisone may be required.

Observing Gelatinous Zooplankton

For those living near the coast, opportunities abound for viewing gelatinous animals. Scuba divers, boaters, kayakers, snorkelers, scientists and naturalists should be able to find many of the animals in this book. It will, however, take many years of searching since some are rare or make only sporadic appearances. With the exception of the exclusively deep-water jellies in this guide, all others at some time or another can be found at or near the surface. As a consequence, even those who cannot fathom the thought of jumping in the water can see most of these animals while staying dry in a boat or looking off a dock. Observations are enhanced if you do have the ability to snorkel or scuba dive since you can then see them in their natural milieu. It is also easier to detect some of the more transparent species from within the water.

Methods for Observing

1) Boating. Perhaps the best way for most people to observe jellies in their natural habitat is to use a small boat, such as an inflatable or Boston Whaler. Sea kayaks are also an excellent way to get a close look at jellies without getting wet

(hopefully). Calm, flat conditions are obviously best, both for your comfort and since jellies tend to head deeper when the surface is rough. A calm day without a significant breeze also enhances the prospects for locating surface slicks. These areas of glassy smooth water, appearing like rivers in the sea when seen from altitude, are usually the best locations for finding large numbers of gelatinous animals. Drifters such as jellies tend to become concentrated in these zones of convergence. Often the most productive slicks are marked by floating kelp, seaweeds and other debris, and surface foam. Slicks will break apart or not form if the breeze is too strong. For this reason it is often best to try to head out in the morning before afternoon winds pick up. Sunny days with the sun positioned behind you are best for seeing jellies, particularly the small, transparent types.

2) Snorkeling. If you have a wetsuit, fins, mask and snorkel, then you can get an even closer look at many types of jellies. The highest densities, particularly of species comprised of smaller individuals, are often within the first five feet below the surface. This makes it quite easy to see many species as long as you are willing to sacrifice being dry. Once again you should search for surface slicks. If there seems to be some jelly action then jump in for a look. Since there is some danger that the boat could be carried away from you by wind or current, it is best to have someone in the boat at all times. Viewing from the comforts of a boat can certainly reveal many jellies. Only by entering the water, however, can you hope to see all that may be present near the surface. Highly transparent species that are invisible from the boat can then be detected by careful searching. Entering the realm of the jellies also enables you to observe their patterns of swimming, escape behaviors, and feeding.

3) Scuba Diving. Although snorkeling is often a productive way to see jellies, scuba diving enables you to expand your range within the water column. This may increase the chances of seeing animals that may be hanging out below snorkeling range. Larger jellies such as *Chrysaora* and *Phacellophora*, for example, are sometimes not visible near the surface but can be found in abundance below 40 feet. Scuba also makes it easier to photograph jellies in their natural habitat. Training and certification are required for scuba diving, and it is more expensive than snorkeling.

Diving with jellies does require some precautions. These animals are drifters and will usually be found beyond kelp beds and other sites that have structures that can be used for orientation. When below the surface while observing or photographing a gelatinous animal, it is often difficult to keep track of where you are in relation to the boat. The anchor line can be used as a reference unless the current is sufficiently strong to rapidly take you out of view. If you find yourself drifting away, then it is best to surface and get close to the boat. Depth should also be closely watched. It is easy to lose track of how deep you may be getting if there is no bottom in view. It should be emphasized that this activity can be extremely dangerous since there is risk of drifting a long distance or suffering air embolism by not paying attention to depth.

The safest way to dive with jellies when the bottom is beyond 40 meters is to use blue-water diving techniques. Researchers who study gelatinous animals have developed methods to safely observe jellies in the open ocean. This requires specialized tethering equipment, in addition to at least three divers and one person in the boat. The system is also somewhat restrictive and makes it difficult to check a number of sites quickly, but for detailed observations of jellies for long periods without concern for where you are drifting, the research technique of blue-water diving is the best method. It is far preferred to the unsafe method of just tying yourself to a long rope attached to the boat. Unless you are associated with a research institution, however, you are unlikely to have a chance to use authentic blue-water diving techniques. (See Heine, J.N., 1986, for detailed information.)

4) Observations From Wharves and Docks. Landlubbers need not despair about never having a chance to see jellies in their natural habitat. Public boat docks and marinas can be great places to see a number of species without the risk of sea sickness. Most will be species commonly associated with nearshore habitats. On occasion, however, you may be blessed by the appearance of some open ocean species that have been swept close to shore. The best times for viewing are either in bright sunlight or at night. Many species tend to swim up into shallower water after the sun sets, and it is also easier to see tiny transparent forms by using a submerged "night light." Another advantage of wharf observation is the ease of quickly checking every day for plankton without the hassles of launching a boat.

5) Public Aquariums. Various types of gelatinous animals have recently become star attractions at a number of public aquariums. This enables anyone without access to a boat or dock the opportunity to view these beautiful animals in a comfortable environment. It is also a good chance for divers and boaters to get a close, detailed look at some of the jellies

they may see in the ocean, in addition to learning a bit about their biology. Aquariums, of course, can display only a limited number of species so you will not be able to see the full spectrum of gelatinous animals. For most people, public aquariums are the closest they will ever get to the planktonic jellies, other than perhaps some gelatinous carcasses washed up on a beach.

6) Scientific Trawls. For over a hundred years, the use of trawl nets has been the primary means for collecting deep-water jellies. This method obviously is a bit rough on these delicate animals and for many types causes complete destruction. There also is little opportunity to observe the natural behavior of trawl-collected jellies. For scientists, naturalists and students who have the opportunity to participate in the deployment and recovery of a midwater trawl, there is plenty that can be observed when the contents of the cod end are brought on to the deck. Many of the deep-water jellies in this book can be recovered relatively intact and placed in a small aquarium for observation. Although they may not survive for long, you can gain some understanding of their appearance and also photograph them. Contrary to popular thought, jellies will not explode when brought up from the depths since they lack gas filled structures that would expand as pressure is reduced.

7) Undersea Vehicles. The last decade has spawned the development of remotely operated vehicles (ROVs), which are gaining in popularity for deep-water observations. ROVs, in addition to more expensive and complicated submersible vehicles that carry people inside and permit direct observations, enable viewing extremely delicate gelatinous animals with minimal disturbance. Unlike fishes and wary invertebrates like squid, most jellies can be approached closely without scaring them away. Obviously most people will find it difficult to come up with the funding necessary to own and operate an undersea vehicle. ROVs do have video capability, however, enabling live viewing by those on shore via fiber optic and microwave links. Increasing numbers of associations between oceanographic and educational institutions will enable more than just a limited group of scientists to visit the jellies of the deep-sea realm.

Collecting Jellies

With bodies that are at least 95% water, gelatinous animals present some difficulties for anyone attempting to collect or display them. Scooping them out of the water with a net followed by placement in a standard fish aquarium is a recipe for disappointment. Using proper procedures with care, however, it is possible to collect even the most delicate specimens with minimal damage.

As with other animals, you must check with Fish and Game authorities in your state before endeavoring to collect gelatinous zooplankton. You may find that regulations are not particularly clear for non-game animals like jellies. In such cases you will need to make the best interpretation from the appropriate officials and published regulations and go from there. With a few exceptions (*Stomolophus*, which is eaten, and *Aequorea*, from which a luminous protein is extracted), jellies in North America are not commercially harvested or endangered so there is no harm to the population by collecting a few.

A container that holds water, such as a bag or plastic jar, is ideal for collecting jellies. Strong bags with rounded bottoms (to prevent trapping jellies in corners of the bag) should be used, if possible. Small gelatinous animals can also be collected in jars or beakers. This can be done from the surface in a boat or at a dock, although it can be a bit easier to get the animals in the bag if you are actually in the water. Scooping a jelly into a jar from a rocking boat is actually a bit more difficult than you may think. Care must be taken to avoid damaging the animal, and you should avoid lifting it out of the water without aqueous support. Sometimes it helps to use your hand (or have a helper) to gently herd the jelly into the collecting container, but just picking up most jellies with your hands or a net will badly damage their fragile bodies.

After the bag or jar is lifted onto the boat or dock, it can be placed in a bucket. It is best to tie the top of the bag with a rubber band after squeezing out all the air. Tops should be placed on jars for the same reason. Removal of the air helps to avoid entrapment of bubbles in the delicate tissue of your gelatinous captives, a situation that can prove fatal. This is particularly important if you anticipate a rough boat ride home. As long as they are kept cool, out of direct sunlight, and not too crowded, the animals should be fine for at least several hours. Adding a small bag of ice or "blue ice" pack to the cooler is usually a good idea. You should avoid mixing species, particularly those that may prey on others in the container. Ideally you should take jellies home or to a laboratory only if you have the facilities to keep them alive (an unlikely prospect for most people). If you are not prepared to maintain a jelly, then the best thing to do is make your observations and take pictures while at the site of collection, followed by gentle release back into the sea.

Maintaining Gelatinous Zooplankton

Jellies have a well-deserved reputation for being difficult to keep very long in captivity. Their delicate tissue and often stringent feeding requirements mean that for the most part only professionals at public aquariums and scientific institutions are likely to have any success at maintaining them for more than a couple of days. Even then many species are beyond any hope of lasting in long-term captivity.

Short-term captivity presents more possibilities. By keeping a jelly for several hours, you have a chance to observe and photograph it. This can be done in a simple 10 or 20 gallon aquarium filled with seawater of the appropriate temperature. The water can be collected at the same time as the jelly. Carefully place it in the aquarium from the collecting jar or bag while keeping the jelly submerged at all times. A spoon or glass rod can be used to gently prod the animal if it tends to rest on the bottom. After several hours of observation it can then be released. Small jellies may be kept in a glass jar or beaker for several days if the water is changed daily and maintained at the appropriate temperature.

Long-term captivity requires the use of specialized aquarium systems that are beyond the scope of most aquarium hobbyists. Public aquariums that display jellies usually use some type of "kreisel" design. Kreisel means "spinning top" in German and refers to the circular flow pattern within the tank. A specialized design for the exit and input of water (needed to create the circulation pattern) enables water motion without sucking the jellies out of the tank. The result is suspension of the jellies off the bottom, and hopefully placement in good viewing position. Many hydromedusae, scyphomedusae and ctenophores thrive with the kreisel design. Other kinds of animals, such as pteropods, heteropods and salps, find difficulties even in a kreisel tank and generally survive for no more than a few days.

Photographing Gelatinous Zooplankton

Jellies present a wealth of photographic opportunities, both in their natural environment or when held captive in an aquarium. Since many are transparent or nearly so, lighting for the image can be a challenge. When done properly, photos of gelatinous animals can be among the most stunning and beautiful of any of the sea's creatures.

Placing jellies in aquaria is a surprisingly effective way to create beautiful photos. With controlled conditions you can reduce some of the problems that cannot be eliminated in the field, such as particles that cause backscatter. The key factor, particularly for highly transparent specimens, is to provide lighting from the top or side. Standard front lighting will only pass directly through the animal and not reflect sufficient light back to the camera and film to form a distinct image. For this reason you will need access to the top or side of the aquarium for placement of the flash unit. This makes it difficult to photograph jellies in a public aquarium if using a flash as the primary source of light.

The photographic chamber should ideally be placed in a darkened room to minimize the chances of reflections from other light sources. You can reduce the problem of reflections even more by wearing dark clothes and gloves. A small light can be placed over the aquarium to provide aid for focusing and composition control. The glass or acrylic surface through which you will be photographing should be absolutely clean. Bubbles and other debris on the inside surface should be wiped some time before the photo session to allow particles a chance to settle out. Suspended particles will reflect light to the camera and show up on your photograph as conspicuous white spots. If the water is excessively cool, you may experience problems with condensation on the outside of the aquarium that must be continuously wiped away with a sponge or squeegee. It is best to use water that is only as cool as necessary to help minimize this problem. Double pane glass or thick acrylic will also reduce condensation. As a background, some sort of untextured, non-reflective dark material, such as a piece of black paper, should be placed behind the photo tank.

The artificial light source (strobe) can be hand-held at the top or side of the aquarium, or supported by some type of bracket. Transparent animals generally require only the use of one strobe unit since their bodies will not create a shadow. More opaque forms may require the use of strobes on two sides of the aquarium (or from the top).

A macro (close-up) lens (50 to 100 mm) with a single-lens-reflex camera (SLR) is generally the ideal photographic system to use. Whether you use automatic or manual exposure control is a matter of choice since excellent results can be obtained either way. Through-the-lens (TTL) exposures work well even with transparent jellies and permit less guesswork for proper exposure. You may wish to experiment with TTL settings to come up with those that work best for you. Manual exposure control will generally require some additional work to determine appropriate aperture settings for particular camera-to-subject distances. For a given distance, you may need to open the aperture a stop or two over what you might normally use for a more reflective subject. Bracketing exposures is advised to ensure that the correct aperture is covered. Focusing is best done manually by moving the camera back and forth until proper focus is achieved. Auto-focus systems tend to have difficulty selecting correct focus on transparent subjects like jellies. A small accessory focusing light permits focus control without contributing light to the image.

Entering the realm of gelatinous animals enables photographs to be made of them undisturbed within their natural environment. This can be done while snorkeling or with scuba, although it's a bit easier to position yourself using scuba. Either housed SLRs or Nikonos systems can be successfully used. A housed camera permits more precise control of positioning of the jelly within the image, but requires excellent buoyancy control by the underwater photographer. Some photographers are more comfortable using a Nikonos camera, which is a viewfinder system that does not allow through-the-lens image viewing. For close-up photography a framer device with extension tubes is usually used with a Nikonos. This may present an advantage while at the surface or when back and forth surge is a problem.

Close approach to jellies is often possible since they generally do not react to your presence as would a fish. Care is still required since some, such as pteropods and heteropods, may swim away when disturbed. Most of the time you will want to be within 2 feet of the jelly unless photographing a large group. The delicate tissue of jellies can be easily damaged or destroyed by careless movement of water or release of air from your regulator. Try to avoid blasting air when you are underneath your photo subject since the large bubbles will send it rapidly to the surface.

A day with calm, clear water with few particles is obviously the best for photographing jellies while diving. Sunlight also helps since it can be used as a backlight to assist in illuminating the often transparent tissue. For relatively opaque jellies like *Chrysaora* or *Pelagia* the sun can be positioned behind the animal to give a nice backlighting effect that helps to show some of the internal structure. Light from a strobe (or 2 strobes, if you prefer) can then be used to fill in the front of the jelly. Highly transparent jellies pass an excessive amount of sunlight through, so you may need to position the sun out of the field of view. Strobe light may then be unnecessary if you can use light from the sun for illumination. By varying the position of the jelly in relation to the background light along with the aperture setting on the camera, you can determine whether the space around the jelly will have a black or more natural blue appearance in the resulting photograph.

Backscatter caused by reflection of light from particles in the water can be a problem. The bright spots that appear on the photograph can be very disturbing. Positioning the strobe so that it is angled between 45° and 90° from a straight-ahead orientation may help to minimize light reflecting back to the camera. This can also help illuminate transparent jellies. The idea is to reduce the amount of light that passes through the area between the jelly and the camera. Depending on whether you use small strobes or larger ones that cover a wider angle of view, you will need to experiment to find your own preferred positioning. You may also try using no strobe lighting and relying entirely on the sun as a light source. With their transparent or translucent tissue, jellies make fine subjects for natural light photography. Even if there are abundant particles in the water, they may not show up on the photograph with this technique.

Photographing large jellies generally requires a wide angle lens such as a 20 mm or 28 mm. This permits close approach while keeping the entire jelly within the boundaries of the image. Getting close is far better than using a normal angle or telephoto lens since it is desirable to minimize as much as possible the distance between the subject and camera when photographing underwater. Close-up photography for smaller jellies can be done with a SLR macro lens (housed camera) or extension tubes and framers (Nikonos). Close-up techniques can also be used on larger jellies to photograph fine details of the animal or some of the creatures that may be hitching a ride.

Photographic Information

David Wrobel's photographs were taken both in natural habitats and in seawater aquariums at the Monterey Bay Aquarium. Photographs in the field were made using a Canon F-1 camera in a housing with a single Ikelite 150 strobe and/or sunlight for illumination. For smaller animals a Canon 50 mm macro lens was used, while larger jellies were captured on film with a Canon 20 mm lens. Aquarium specimens were photographed in clean kreisel tanks or a 5 gallon acrylic tank with black backgrounds. Single strobe illumination from the side or top of the tank provided the source of light. Either a Canon

F-1 camera with Canon 50 mm macro lens and Vivitar 283 flash (manual exposure control) or a Canon A2 camera with Canon USM 50 mm macro lens and Canon Speedlite 430 EZ flash (TTL exposure control) was used. Fujichrome Velvia, Kodachrome 64 and Kodak Ektachrome 100 slide films were used.

Claudia Mills' photographs were taken primarily in the laboratory, using carefully hand-dipped undamaged specimens. The jellies were gently placed in straight-sided homemade glass aquaria of 50 ml to 15 liter volumes filled with filtered, chilled seawater, and with a black background. The animals were never handled directly, but were moved using one of several glass or plastic bowls, long-bore pipettes, or beakers. Most of the photographs were taken in a walk-in coldroom at the Friday Harbor Laboratories (University of Washington) where the animals could remain at their correct seawater temperature for many hours or even a couple of days. Kodachrome 64 film was usually used, with a Nikon F3 camera, Nikon micro-Nikkor 105 mm lens (sometimes with a 1:1 extension ring), and a single Nikon SB-17 flash held in a clamp overhead, or occasionally to the side, and metered with a TTL cable. Black and white photographs were taken using the same set-up with Kodak Pan-X, Plus-X, Tech Pan, or T-Max 100 films.

Glossary

Aboral Side or end opposite the mouth of a medusa or ctenophore
Apical sense organ Structure at the aboral end of ctenophores (and some kinds of larvae) which is used in orientation
Atrial aperture The posterior opening in salps and doliolids through which water leaves the body
Atrial chamber In thaliaceans, the compartment within the body into which the cloaca opens
Auricle Ciliated appendage in lobate ctenophores that aids prey capture
Basal bulb Swelling at the base of a tentacle of a hydromedusa, where it connects to the bell margin
Bell The umbrella-shaped, pulsing structure of medusae and siphonophores used for locomotion
Blastozooid Early buds, produced asexually, which then develop into the sexual generation
Bracts Transparent leaf-like gelatinous protective zooids that cover other parts along a siphonophore stem
Branchial bars Rigid structure in salps that supports the gills (also known as gill bars)
Branchial basket In thaliaceans, the structure formed by the gills and gill apertures through which water passes
Branchial openings Apertures in thaliacean gills through which water passes
Buccal cavity Oral cavity of molluscs
Buccal cones Eversible oral tentacles of some gymnosome pteropods used in capturing prey
Cadophore Linked embryos forming a tail-like structure on doliolid oozooids
Calyx The wide-opening main portion of the body of a stauromedusa, above the stem
Capitate Describes a hydrozoan tentacle that has a distinctly round, swollen tip
Centripetal canals Blind canals which arise from the ring canal in hydromedusae, growing upward toward the bell apex, but usually not reaching it
Chromatophores Pigment-filled sacs in the epidermis that are used in color changes
Colloblasts Specialized sticky cells on ctenophore tentacles which are used for prey capture
Comb rows Unique ctenophore structures formed by series of plates of fused cilia, used for locomotion
Concretion Tiny calcareous grain within a statocyst (also called otolith)
Cormidium A repeated segment of a siphonophore, each with a gastrozooid, gonozooid and dactylozooid; cormidia that break free are called eudoxids
Ctenidium The gill of a mollusc
Dactylozooid Nematocyst-bearing tentacle-like zooid of a siphonophore or hydroid used for prey capture and defense
Demersal zooplankton Species that sink to the bottom by day and rise up into the water column at night
Dextral coiling Shell of a gastropod coiled in a clockwise direction
Diatoms Unicellular phytoplankton, usually with cell walls made of silica
Dinoflagellates Unicellular algal-like planktonic organisms with a pair of flagella
Ectoparasite A parasitic organism that lives on the exterior surface of the host
Endostyle Mucus-producing structure associated with the pharynx in tunicates
Ephyra Post-larval, youngest medusa stage of a scyphomedusa
Epifauna Animals that live on a substrate
Epipelagic zone Oceanic area from the surface down to about 200 meters

Eudoxid A small group of siphonophore zooids (see cormidium) that break free from the stem and live independently, which serve to disperse gametes

Euphausiids Shrimplike crustaceans that associate in large planktonic groups

Euphotic zone Layer of the ocean from the surface to the maximum depth that receives sufficient light for photosynthesis

Exumbrella The outside surface of the bell of cnidarian medusae

Filiform Describes a tentacle of a hydroid or jellyfish that is simple and unornamented

Foraminifera A group of protozoa that possess a calcareous exoskeleton

Gastric cavity Sac where food passes after entering the mouth and begins digestion in cnidaria and ctenophores (also called manubrium, stomodaeum, or pharynx)

Gastric peduncle Gelatinous bulge protruding into the roof of the subumbrella of a jellyfish, bearing the stomach or manubrium

Gastrozooids Individuals in a siphonophore that are primarily responsible for digesting captured prey

Gonozooids Individuals that produce eggs and/or sperm

Hermaphrodite An individual animal that possesses both male and female gonads at some point in the life cycle

Holoplanktonic Animals that complete their entire life cycle in the water column

Hook sacs Gymnosome pteropod structures containing hooks used in prey capture

House External mucous sac formed by larvaceans for capturing food

Interradial In describing the symmetry of hydromedusae, the four planes meeting at right angles at the summit of the umbrella between the four perradii, that is, equally spaced between the four radial canals

Macrocilia Modified cilia in the mouth of beroid ctenophores which aid capturing gelatinous prey

Mantle An extension of the body wall of molluscs that forms the shell and lines the largest shell whorl

Mantle cavity The space in molluscs between the lining of the mantle and the rest of the body

Manubrium Variously shaped, pendant, subumbrellar gastrovascular cavity (stomach) in medusae, bearing a terminal mouth

Marginal vesicle Statocyst in a small pit at the bell margin of hydromedusae

Medusa The post-larval planktonic free-swimming phase of scyphozoan and hydrozoan cnidarians (a jelly)

Mesentery Connecting membrane between radial canal and manubrium of certain hydromedusae

Mesoglea The gelatinous material (jelly) making up most of the volume of the body of cnidarian jellies and ctenophores

Mesopelagic zone Oceanic layer from the bottom of the epipelagic zone (200 m) or transition zone (400 m) to about 1000 meters depth

Metachronal Beating of cilia one after another in a line, forming a wave

Nectophore Swimming bell of some siphonophores

Nekton Pelagic animals with relatively strong swimming abilities (faster than ocean currents carry them), such as fish and most squid

Lappets Marginal flaps on the bell of scyphomedusae

Nematocysts Capsules with an eversible thread held within specialized cells in the tentacles and other areas of cnidarians, they carry toxins and cause the "sting" (also called cnidocysts)

Neritic Coastal waters with the bottom less than about 200 meters depth

Nomen nudum Scientific name that is used without prior publication of a description, and is thus invalid

Notochord Primitive type of skeletal rod, found at some stage in chordates (including tunicates)

Nurse Late stage doliolid which carries blastozooids

Oceanic Offshore waters beyond the continental slope

Ocellus Pigmented, light-sensitive spot on the tentacle base or bell margin of certain medusae

Oozooid Individual thaliacean that develops from a sexually-produced egg

Operculum Chitinous structure on the foot of molluscs that seals the aperture of the shell

Oral aperture Anterior opening of salps through which water is pumped into the body

Oral arms Long frilly extensions of the mouth of scyphomedusae

Ovotestis Gonad of hermaphroditic animal which produces both eggs and sperm

Peduncle Stalk by which aggregate salps are attached to each other (see also gastric peduncle)

Pelagic Of the open sea

Perradial In describing the symmetry of hydromedusae, the four planes meeting at right angles at the summit of the umbrella in which the four radial canals lie

Phorozooid Stage in doliolids which functions primarily for locomotion (sexual gonozooids attached)

Phytoplankton Diatoms, dinoflagellates and other microscopic planktonic algae

Plankton Plants and animals with relatively limited swimming ability that are at the mercy of currents

Planula The microscopic, ciliated larva of a cnidarian

Pneumatophore Gas-filled sac at the uppermost end of some siphonophores

Polyp The sessile life history stage of many hydrozoans and scyphozoans

Proboscis An extendible organ used in feeding, usually with a terminal mouth

Protandrous hermaphrodite An hermaphrodite that starts as a male, and later changes into a female

Pseudoconch A thin, non-calcified shell-like structure present in some thecosome pteropods

Pseudofeces Particles rejected before ingestion by ciliary mucous feeders

Radial canals Circulatory tubes in medusae that connect the gastric cavity with the marginal ring canal

Radula A rasping feeding organ with rows of chitinous teeth used by some molluscs

Reciprocal fertilization The fertilization of each other's eggs by both members of an hermaphroditic pair

Rhinophores Sensory structures present on the head region of many nudibranchs

Rhopalium Sensory structure in the margin of the bell of scyphomedusae

Ring canal Circular, tubular canal running around the bell margin of a medusa, connected to the gastric cavity via the radial canals

Scyphistoma Polyp of scyphozoans

Side branches Also known as tentilla, the filaments that branch off the main tentacle axis in cydippid ctenophores

Sinistral coiling Counterclockwise spiraling of gastropod shells

Slope water Waters overlying the continental slope

Spermatophore A sperm package that can be transferred from a male to another individual for fertilization

Statocyst Sense organ used for balance and orientation

Stolon Steadily elongating, asexually produced chain of salps formed by solitary individuals; also, the tubular structures between polyps in a colonial hydroid, usually attached to the substrate

Strobilation Process by which transverse divisions of a scyphistoma results in the release of ephyrae

Subumbrella The inside surface of the bell of cnidarian medusae, bordered by the bell margin

Tentacles Long, thin food gathering structures in cnidarians and ctenophores

Tentacle bulbs In hydromedusae, the swollen tentacle bases at the bell margin; in ctenophores, the paired, swollen structures at the bases of the tentacles, usually near the pharynx

Tentacle sheaths The paired funnel-like tubes through which the tentacles of cydippid ctenophores exit the body

Tentilla The side branches of a ctenophore tentacle (singular tentillum)

Tintinnids Shelled, planktonic ciliate protozoans

Tunic Relatively firm body wall of tunicates, made of a cellulose compound (also known as the test)

Umbrella The bell-shaped swimming structure of medusae

Veliger Larval stage of many molluscs

Velum Thin, muscular flap-like structure surrounding the margin of the bell in a hydromedusa; also the ciliated, feeding and swimming organ of the veliger larvae of molluscs

Visceral mass Gut and associated structures of pelagic molluscs and tunicates

Wing plate Fused, paired wings of pseudothecosome pteropods used for locomotion

Zooid A single individual in a colony

Zooplankton Animals that drift in the ocean

Guide to Body Parts

The following photographic illustrations, featuring representative species among the major gelatinous animal groups, should be used as general guides to the main body parts of the animals featured in this book. These illustrations should not be considered as comprehensive treatments of gelatinous animal structure. Species within a group, such as hydromedusae, may vary somewhat from the body plans presented here.

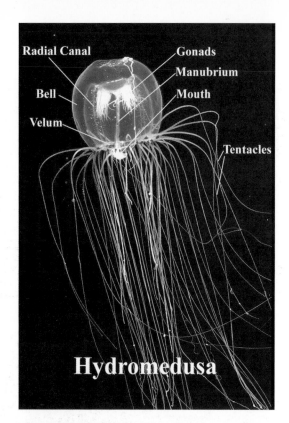

Hydromedusa

Radial Canal
Gonads
Manubrium
Bell
Mouth
Velum
Tentacles

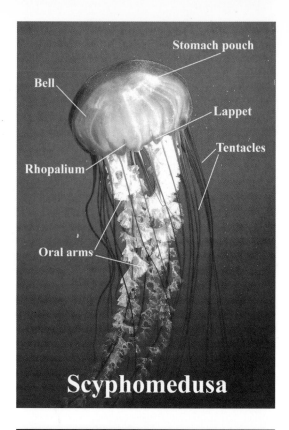

Scyphomedusa

Stomach pouch
Bell
Lappet
Rhopalium
Tentacles
Oral arms

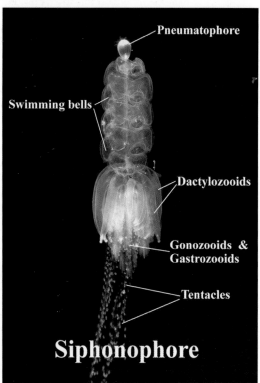

Siphonophore

Pneumatophore
Swimming bells
Dactylozooids
Gonozooids & Gastrozooids
Tentacles

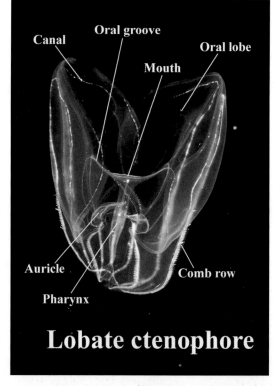

Lobate ctenophore

Canal
Oral groove
Oral lobe
Mouth
Auricle
Comb row
Pharynx

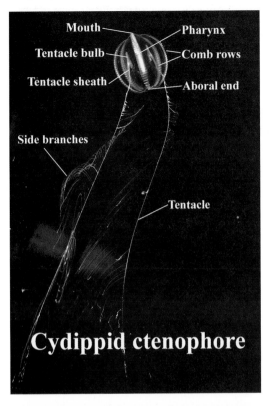

Cydippid ctenophore

Mouth
Pharynx
Tentacle bulb
Comb rows
Tentacle sheath
Aboral end
Side branches
Tentacle

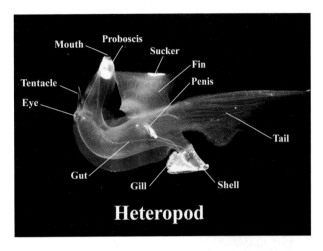

Heteropod

Mouth
Proboscis
Sucker
Fin
Tentacle
Penis
Eye
Tail
Gut
Gill
Shell

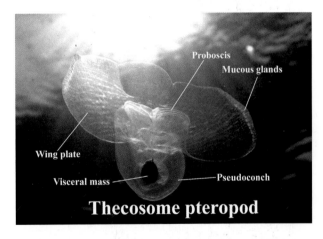

Thecosome pteropod

Proboscis
Mucous glands
Wing plate
Visceral mass
Pseudoconch

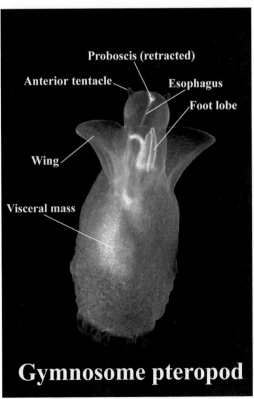

Gymnosome pteropod

Proboscis (retracted)
Anterior tentacle
Esophagus
Foot lobe
Wing
Visceral mass

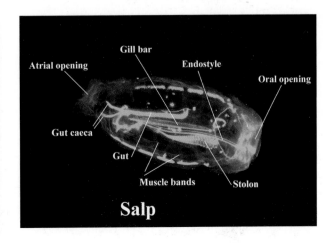

Salp

Gill bar
Endostyle
Atrial opening
Oral opening
Gut caeca
Gut
Muscle bands
Stolon

Hitchhikers on Jellies

For creatures of the open sea realm, there are precious few protective sites. Many gelatinous animals serve as convenient traveling homes or resting places for a variety of other creatures. Certain types of larval fish and crustaceans are the primary users of this resource. Careful observation of gelatinous zooplankton will often reveal the presence of one or more hitchhikers.

Hitchhiking serves a number of purposes. Some larval or juvenile animals use their gelatinous host as a platform for development to adulthood. Other species may spend their entire lives on a jelly after settling down. Juvenile fishes, such as the medusafish (*Icichthys lockingtoni*), Pacific butterfish (*Peprilus simillimus*), and walleye pollock (*Theragra chalcogramma*) often lurk in the vicinity of a large scyphomedusa. When danger approaches, they dive into the protective confines of the bell. Medusafish are even occasionally seen inside large salps. Crabs, such as the slender crab (*Cancer gracilis*), often spend their formative months in association with a jellyfish before assuming a benthic existence. *Pelagia colorata* seem particularly favored by these crabs. Most hitchhikers grab food that the host has collected, but they may also consume host tissue. For this reason the association can be somewhat deleterious to the gelatinous host. An association that is certainly unfavorable to the host is that between the larval sea anemone, *Peachia quinquecapitata*, and certain hydromedusae. Young anemone larvae are ingested by the medusae, and feed on the gonads and stomach of their hapless jellyfish hosts before dropping off and assuming a more typical benthic lifestyle as adults.

A large number of amphipods in the family Hyperiidae are associated with gelatinous animals. Medusae, siphonophores, ctenophores, pteropods and salps all serve as homes for hyperiids. Often an amphipod will excavate a protective pit in the tissue of the host, or may be embedded deeper inside the animal. Some living amphipods can even be found inside the stomachs of hydromedusae. Females of one hyperiid amphipod, *Phronima*, actually take over the tests of certain pelagic tunicates and swim while covered in their modified "barrel." *Phronima* broods eggs within the barrel, and the hatchlings then consume their home before searching for more salp victims. Certain salps are also used by males of the epipelagic octopus, *Ocythoe tuberculata*. The octopus uses jet propulsion to swim, even while inside its gelatinous home.

Reproduction in Pelagic Cnidarians

Most hydrozoans and scyphozoans have a sessile, attached polyp stage, unlike the holopelagic ctenophores, pteropods, heteropods, and pelagic tunicates. This complex life cycle, with both attached and free-swimming stages, is a sort of alternation of generations. Most medusae free-spawn eggs and sperm individually into the water, where close proximity is the primary determinant of which eggs manage to be fertilized. (A small number of medusae retain the eggs, which are internally fertilized by freely spawned sperm, and then brooded for awhile.) Typically, ciliated planulae are produced from fertilized eggs released by the sexual medusae (there are separate male and female sexes). These settle on the appropriate substrate and soon form individual polyps. Polyps may reproduce asexually to create permanent colonies that can bud and release indefinite numbers of free-swimming medusae. Attached polyps of the orders Leptomedusae and Anthomedusae are often as prominent as the medusae; in the scyphozoans the medusae generally are far more conspicuous than their sessile polyps.

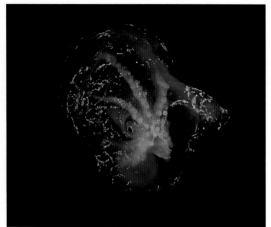

Ocythoe tuberculata   B. Heller
Epipelagic octopus in salp

Chrysaora fuscescens with commensal butterfish
(*Peprilus simillimus*)

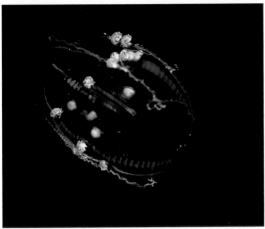

Hormiphora with hyperiid amphipods

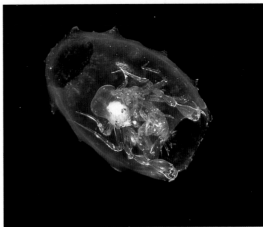

Phronima sedentaria
Hyperiid amphipod in dead salp

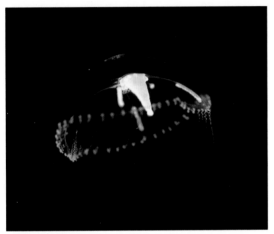

Clytia gregaria with parasitic
larval sea anemone (*Peachia*)

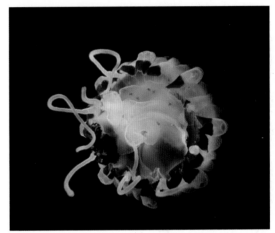

Pelagia colorata with commensal juvenile crabs

Among the hydrozoans, Trachymedusae and Narcomedusae (which are usually open ocean and deep-sea dwellers) lack polyps and have a free-swimming larval stage that develops directly into another medusa. Anthomedusae, Leptomedusae and Limnomedusae have retained the polyp stage in their life cycles. Although we typically think of polyps as occupying benthic habitats, open ocean species in these groups have evolved specialized habitats from available substrata including pteropod shells, clumps of filamentous algae, and even fish skin. Hydroids (another term for the polyp form of hydrozoans) can produce either medusae or more polyps by asexual budding for indefinite periods of time. The individual polyps are usually small and inconspicuous, and depending on the species, may form tightly associated colonies that can have several different kinds of specialized zooids. Some species have hydroid colonies that are relatively large and easily visible; many species of hydroids produce no free-swimming medusae at all. Since the hydroids are drastically different in appearance from their medusae, it can be difficult to determine what species of medusa is produced by a particular type of polyp. This has resulted in numerous cases in which a hydroid and the medusa it produces have been described with different scientific names. Other species which presumably have a polyp stage are well known as medusae, but have not been linked with any type of known hydroid. A goal of scientists who work on these groups is to eventually link all polyp-medusa pairs under the same names.

Scyphozoan polyps, or scyphistomae, of the orders Semaeostomeae and Rhizostomeae, are generally relatively large (typically several millimeters long). Although they often clump together, scyphistomae live as individuals rather than in unified, polymorphic colonies. The polyps of some medusae in the order Coronatae, however, do form colonies, with a chitinous covering, not unlike certain hydroids. Scyphozoan polyps can reproduce asexually by a variety of means to form new scyphistomae. When environmental conditions are appropriate, a scyphistoma develops horizontal constrictions, somewhat similar in appearance to a stack of plates. This process, known as strobilation, is highly seasonal in many species, and results in the sequential release of many 8-lobed, free-swimming ephyrae. The ephyrae gradually assume the adult body form and eventually reach maturity as separate male and female individuals.

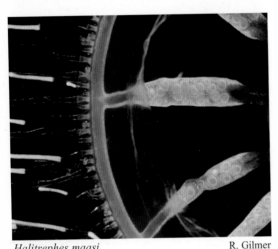

Halitrephes maasi   R. Gilmer
Hydromedusa with eggs in the gonads

Sarsia sp.
Hydroids with medusa buds

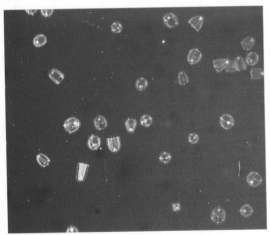

Velella velella
Newly released medusae

Phacellophora camtschatica
Strobilating scyphistoma

Phylum Cnidaria / Class Hydrozoa

Of all the gelatinous animal groups featured in this book, the Hydrozoa exhibit the most bewildering variety of life histories. In addition to the hydrozoan jellies and their usually-benthic hydroids, this diverse class also includes the hydrocorals, which form massive benthic colonies using calcium carbonate as a skeletal material. Many of the tiniest gelatinous animals that you will encounter are hydrozoans, with some only a couple of millimeters across. Other species are among the largest gelatinous animals of the sea, with a few siphonophores exceeding 40 meters in length. The most potent stingers on the west coast are siphonophores, such as *Physalia*, the Portuguese man-of-war. Whereas other groups

Aurelia aurita
Strobilating scyphistomae

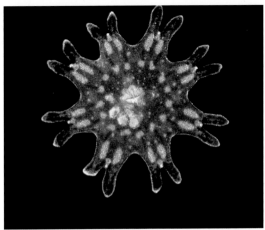

Pelagia colorata
Ephyra

of gelatinous animals are exclusively marine, a few hydrozoan species have adapted to the rigors of living in fresh water or very low salinity habitats. Hydrozoan life histories range from those species possessing an entirely free-swimming (holoplanktonic) life style to those that live exclusively as an attached polyp form, with many species possessing both medusoid and polypoid stages.

Subclass Hydromedusae

The five hydrozoan orders that represent west coast hydromedusae generally include species that are relatively small or moderate in size compared to most scyphomedusae. They typically have fragile, transparent tissue, with many being difficult to see in the water. Most are easily overlooked by the casual observer, in contrast to conspicuous scyphomedusan giants like *Chrysaora* and *Pelagia*, which are hard to miss. The smallest species are perhaps most likely to be encountered when sorting plankton tows. Hydromedusae possess a thin, muscular flap of tissue known as the velum, which is attached to the inner margin of the bell. Presence of the ring-like velum is a key feature distinguishing this group from the Scyphozoa, and is important for propulsion. The mesoglea of hydromedusae lacks living cells, in contrast to that of scyphomedusae. Another distinguishing characteristic is the gastric cavity, which in Hydrozoa is a simple sac lacking divisions into separate pouches.

Hydromedusae can be very difficult to identify to species, even by specialists, and often require use of a microscope. Species are differentiated by minor variations in characteristics such as shape and size of the bell, position of gonads, number of radial canals, presence or absence of eyespots and/or statocysts, number of tentacles, and size and shape of the manubrium. Classification is further complicated by the difficulty in linking medusae with their particular polyp stage. Although medusae and polyps should be given the same name (the oldest correct one) once their complete life cycle is known, you may find instances in the modern literature in which the scientific name still in use for a medusa is different than that of its corresponding polyp.

If you observe a jelly and determine that it is a hydromedusa, then try to place it within one of the five orders. Most of the nearshore species will be either Anthomedusae or Leptomedusae. The easiest first guess in distinguishing these orders is given by the shape of the bell: Anthomedusae usually have a tall bell and most Leptomedusae often have a flattened bell. Narcomedusae and Trachymedusae are more common in oceanic and deep-water habitats. Narcomedusae have a thick lens-like central bell region and unique, relatively rigid tentacles that are often held up above the bell rather than trailing behind as in the typical jellyfish posture. Trachymedusae usually have large numbers of tentacles that break off when handled roughly (as in net collection) and very substantial swimming musculature. This book features most of the relatively common hydromedusae, along with some that are rarely observed. We have not included all of the species that occur along the west coast, so you may encounter some hydromedusae that are not shown.

Hydromedusae swim using jet propulsion with a very thin muscle layer on the inside of the bell (the subumbrella). Many species possess ocelli (light sensitive eyespots) that seemingly enable the medusa to move toward or away from light, but some species without ocelli also demonstrate light-sensitive behavior such as migration to deeper water during the day and rising closer to the surface at night. Statocysts, or balance organs, are another feature of certain species and help to determine proper orientation of the bell. Nearly all hydromedusae possess tentacles which are used for capturing zooplankton prey, including copepods and other small crustaceans, larval fish, comb jellies, other hydromedusae, appendicularia, various larval invertebrates, and fish eggs and juveniles. With their transparent bells, it is often easy to determine what types of prey a hydromedusa has eaten by visual inspection of the gut. Compared to siphonophores and scyphozoan jellies, the nematocysts of most hydromedusae are relatively mild and do not present much of a threat to people. Nevertheless, it is still prudent to avoid handling these jellies since a few species, such as the Russian population of *Gonionemus vertens*, have a very potent sting.

Order Anthomedusae

Anthomedusae are a group of medusae usually characterized by a bell as tall or taller than wide, with gonads located on the stomach (manubrium) walls, and without statocysts; some Anthomedusae have ocelli. Anthomedusae all have a hydroid portion of their life cycle–the hydroids are called "athecate," since the polyps are not enclosed in a chitinized cup, or theca. Polyps can be solitary or colonial, and may have stems and stolons covered by a chitinized skeleton. Many athecate hydroids also classified in this order do not produce medusae, but we do not include any of those species here. Anthomedusae (and Leptomedusae) might be considered in the crudest sense to be swimming and feeding gonads of their hydroids, whose life in the plankton aids in dispersal of their free-spawned gametes.

Order Leptomedusae

Leptomedusae are a group of medusae usually characterized by a bell as wide or usually wider than tall and with gonads located on the radial canals; most have statocysts and a few species also have ocelli. Leptomedusae all have a hydroid portion of their life cycle–the hydroids are called "thecate," since the polyps are enclosed in a protective chitinized cup, or theca. The hydroids of Leptomedusae are usually colonial and the stems and stolons of the colonies are also covered by this chitinous material. The majority of thecate hydroids in this order do not produce medusae, but we do not include any of those species here. The largest hydromedusae are members of the Leptomedusae. Many are bioluminescent around the bell margin; this is best seen by taking a healthy specimen into a dark room, letting one's eyes adjust to the darkness for a minute or two and then gently agitating the jellyfish while watching carefully. The bioluminescence usually appears to be green, as it is emitted in association with a green fluorescent protein in many species.

Order Limnomedusae

Limnomedusae are a small, but diverse group of jellyfishes so-named because the first representatives placed in this group live in fresh water; a few marine species have since been added. Limnomedusae are usually about as tall as they are broad and have gonads either on the stomach wall or on the radial canals. The medusa tentacles are usually ornamented with clasps or rings of nematocysts. Their athecate polyps are generally very small and inconspicuous, and are often solitary rather than colonial; many can bud to asexually reproduce more polyps. As their biology has become better known, some species from this group have been moved into the Anthomedusae.

Order Narcomedusae

Narcomedusae are characterized by a thick lens-like umbrella with stiff central mass of jelly and thin scalloped edges; heavy solid tentacles come off the bell well above the margin and in healthy specimens are often held curved up above the bell. The stomach is broad and circular. Narcomedusae lack a true hydroid stage and have a holoplanktonic life cycle. Some species have a parasitic larval stage that lives on the surface of other species of jellyfishes and in some cases can bud additional individuals from a hydroid-like stolon. Most Narcomedusae are oceanic or deep-sea in habitat, although representatives of the genus *Solmaris* may occasionally be very abundant in warm to warm-temperate nearshore waters.

Order Trachymedusae

Trachymedusae are a group of medusae usually characterized by a bell that is either hemispherical or taller than wide, with gonads located on the radial canals, with statocysts at the bell margin, and without ocelli. All Trachymedusae whose life cycles are known are holoplanktonic. The male and female medusae free-spawn eggs and sperm which are fertilized in the sea, with embryos developing directly into another medusa. The tentacles of Trachymedusae are usually as long or longer than the bell is tall, but they are fragile and usually break off when collected in nets, leaving very short, uniform stubs which some authors have assumed to be the entire tentacles. Most Trachymedusae are oceanic or deep water in habitat, but *Aglantha* is found in both coastal and open waters. The heavy swimming musculature of the subumbrella and velum of most Trachymedusae is birefringent in certain light conditions.

Order Anthomedusae
Family Bougainvilliidae

1. *Bougainvillia* spp.

Identification: Bell to about 10 mm high and wide; Manubrium with 4 very distinctive dichotomously-branched oral tentacles that in some cases fill much of the subumbrellar space. Gonads on the manubrium. With 4 radial canals and 4 tentacle bulbs; each tentacle bulb bears many similar tentacles (up to 60), each with its own dark ocellus. Umbrella transparent, with pinkish to brownish manubrium. *Natural History*: Fairly short-lived, occurring in late spring or summer and feeding especially on barnacle nauplii. Preyed on by *Aequorea* and other medusivorous hydromedusae, which may cause the demise of the *Bougainvillia* population. *Remarks*: Several species of *Bougainvillia* have been recorded in the coastal waters of Washington, British Columbia, and Alaska, including *B. britannica*, *B. multitentaculata*, *B. principis* and *B. superciliaris*. The related *Chiarella centripetalis* is found in the Gulf of California.

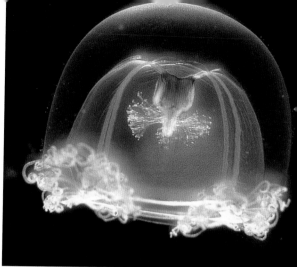

D. Lickey

Family Calycopsidae

2. *Bythotiara stilbosa*

Identification: Up to 5.4 mm in width and 4.8 mm in bell height; bell studded with single nematocysts over the exumbrellar surface. Stomach surrounded by 8 gonads; mouth cruciform, with lightly ruffled lips; without gastric peduncle. With 4 unbranched radial canals; without statocysts; velum narrow. With 4 tentacles having swollen tips. Umbrella transparent, with red-brown pigment in the gonad and tentacle bases, and orange and red pigment on the knobbed tentacle tips. *Natural History*: Although most members of this family are considered to be oceanic, newly released medusae of this species were collected at a marina in a shallow harbor. The undescribed polyp of this species is assumed to be present in Bodega Harbor. *Range and Habitat*: Known only from Bodega Harbor, central California.

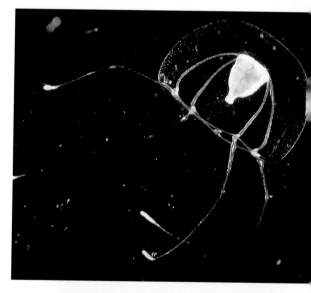

3. *Calycopsis simulans*

Identification: Bell to about 40 mm tall; manubrium rather small, with 4 recurved lips; with 4 tightly folded gonads on the manubrium. With 12 or more large tentacles and a few small ones. With 4 radial canals arising from the corners of the manubrium and an additional 12 or more centripetal canals that arise from the margin and may meet the base of the manubrium, but not at the corners. Bell transparent, manubrium deep brick-red, with orangish pigment near the tips of the large tapering tentacles. *Range and Habitat*: Eastern Tropical Pacific to the Bering Sea, near the surface. Uncommon.

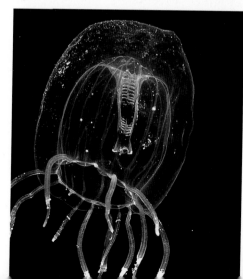

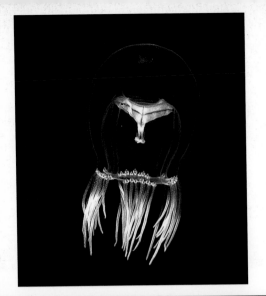

Family Halimedusidae

4. *Halimedusa typus*

Identification: Up to 16 mm tall and 13 mm wide; manubrium on a broad low peduncle, cruciform, with a small quadrate mouth. Gonads on the manubrium, extending out towards the 4 straight, smooth radial canals. With 4 perradial tentacles and 4 interradial groups of up to 10-11 tentacles, all with red or purple-black ocelli on their basal bulbs. Umbrella transparent, with a whitish gonad usually with a dark line running out each perradial gonad. *Natural History*: Nothing seems to be known about the natural history of this coastal medusa. Solitary polyps less than 0.5 mm tall have been raised in the laboratory. *Range and Habitat*: Originally described from Amphitrite Point, Vancouver Island. Abundant in Humboldt Bay, California. Also known from Bodega Harbor, California and Yaquina Bay, Oregon. Present summer and autumn.

Family Pandeidae

5. *Amphinema platyhedos*

Identification: Up to 5.5 mm wide, with a pointed apical projection. Manubrium large, covered by 4 horseshoe-shaped swollen, but not folded gonads; mouth with 4 slightly recurved lips. With 4 radial canals, but with only 2 very large opposite tentacles having large, broad basal bulbs and with 26 small tentacles without basal bulbs. Umbrella transparent, with yellow, tan or pale orange manubrium, lips, gonad and large tentacles. *Natural History*: In situ, this species is usually seen with its two large tentacles extended directly out forming a straight line from the margin of the bell. Eats siphonophores in the laboratory. *Range and Habitat*: Known from deep water in British Columbia and southern California. Similar specimens have been seen in surface waters off Santa Barbara and Monterey (pictured here). It is not entirely clear that all are the same species.

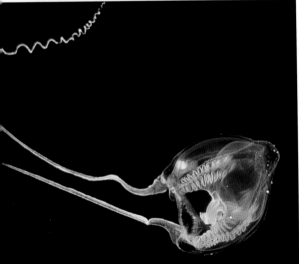

6. *Annatiara affinis*

Identification: Up to 23 mm in diameter, jelly moderately thick; very fragile. Exumbrella with distinctive long nematocyst tracks running upward from every tentacle base. Manubrium with 4 lightly ruffled lips; without gastric peduncle. Gonads each with up to 20 irregular vertical folds, covering the entire stomach which extends some distance out along the 4 broad, straight radial canals. With up to 44 tentacles of different sizes. Tentacle bulbs laterally compressed and clasp the bell margin; each has a red ocellus. Umbrella transparent, with white gonads and pale yellow tentacle bases. *Range and Habitat*: Cosmopolitan; rarely collected and usually in bad condition. Usually considered a deep water species, but several specimens have been found near the surface in Monterey Bay.

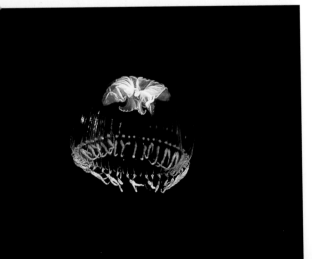

7. *Catablema nodulosa*

Identification: Up to about 20 mm wide, with large, rounded apical gelatinous process. Stomach with broad basal attachment continuing a short distance down radial canals; with frilly lips. Gonads on manubrium, with perpendicular to vertical folds. With 4 broad radial canals with uneven edges. With 8, 16, or 32 tentacles, and 1–3 marginal bulbs between adjoining tentacles; without ocelli. Gonads and stomach salmon pink to golden-brown. *Natural History*: Sometimes abundant in late spring in the Puget Sound–Strait of Georgia area, where it is a voracious predator on other hydromedusae and ctenophores. Hydroid not known. *Range and Habitat*: Puget Sound and British Columbia to Aleutian Islands and Bering Sea; upper 100 meters. *Remarks*: The larger species *C. multicirrata*, with up to 320 tentacles, is found British Columbia to Alaska.

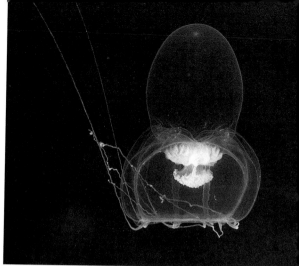

D. Lickey

8. *Halitholus* spp.

Identification: Bell to about 12 mm tall, with a rounded to pointed gelatinous apical projection. Manubrium large and attached to the radial canals only from the top (without "mesenteries"); with large mouth. Gonads on the manubrium, usually horseshoe-shaped and with folds. With 4 radial canals, usually with smooth edges. With 8–16 tentacles with laterally-compressed bulbs; with ocelli. Bell typically transparent, with some reddish to yellowish to brownish color in the manubrium. *Natural History*: Small and inconspicuous; often passively hanging upside down, suspended below the long tentacles. One *Halitholus* hydroid colony was found in the egg mass of a *Cancer gracilis* in Friday Harbor. *Range and Habitat*: Coastal in temperate and boreal waters. *Remarks*: Species of *Halitholus* have been reported from Washington, British Columbian and Alaskan waters, including *H. pauper*, *H. cirratus*, and 2 undescribed species.

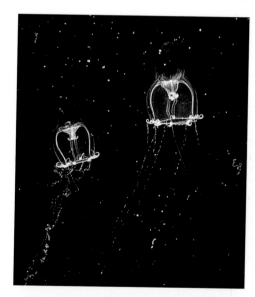

9. *Leuckartiara* spp.

Identification: Bell usually 8–20 mm tall, with gelatinous apical projection. Manubrium large and broad, attached partially to radial canals by membranes called "mesenteries"; large frilly mouth. Gonads on manubrium, usually horseshoe-shaped, with folds. With 4 radial canals, usually broad and ribbonlike, sometimes with jagged edges. With 12–48 tentacles with laterally-compressed bulbs; with ocelli. Bell transparent, with some reddish to yellowish to brownish color in the manubrium and tentacle bases. *Natural History*: These medusae probably live a month or two and prey on other gelatinous zooplankton. The tentacles can be extended many body lengths. The hydroids are small, slightly-branched colonies. *Range and Habitat*: Coastal, late spring and summer.

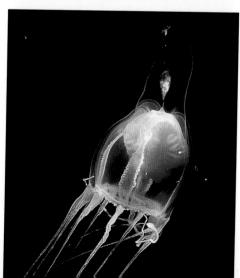

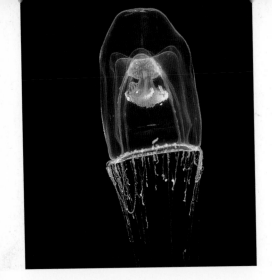

10. *Neoturris breviconis*

Identification: Bell to 45 mm high; gelatinous apical projection low and conical. Manubrium broad, lips with fine frills at the margin. Gonads on manubrium, horseshoe-shaped, with horizontal to oblique folds on the sides and often with small pits near the center. With 4 broad radial canals with jagged edges. With 100 or more densely crowded tentacles, each with a laterally-compressed bulb; without ocelli. Bell transparent, gonads and manubrium orange, pink or red. *Range and Habitat*: Occurs in both North Atlantic and North Pacific in boreal and arctic waters. Seen occasionally in inland waters of Washington and British Columbia in the late spring; also known from Monterey Bay and the Bering Sea. Likely to occur along most of the Alaskan coastline. Hydroid not known.

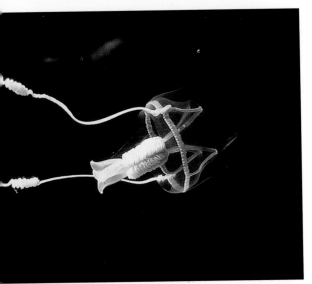

11. *Stomotoca atra*

Identification: Bell to about 25 mm high and wide; stomach large, suspended by a broad gelatinous peduncle, with 4 recurved lips. Gonads on the manubrium, with regular transverse folds. With 2 large opposite tentacles and about 80 rudimentary marginal tentacles. Gonads dark brown, tentacles light brown to tan. *Natural History*: An active swimmer alternating bouts of swimming and sinking, it feeds primarily on other hydromedusae, especially *Clytia gregaria*. Has a strong "crumple" response when touched, which causes it to stop swimming and sink. A species of *Stomotoca* in the Caribbean has a hydroid that lives on eel larvae. *Range and Habitat*: Santa Barbara to the Bering Sea; common in summer in Puget Sound, British Columbia and Alaska. A similar medusa given the same name has been collected in the tropical Pacific waters of Papua New Guinea.

D. Lickey

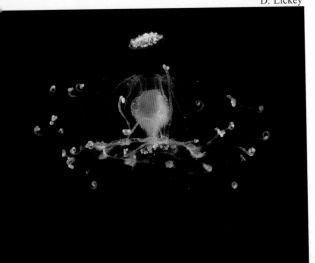

Family Rathkeidae

12. *Rathkea octopunctata*

Identification: Bell to about 4 mm high; manubrium short, on a gelatinous peduncle. With medusa buds developing on outer walls of the stomach on younger specimens, this space later occupied by developing gonads. Mouth with 4 distinct lips, each characterized by a pair of separated nematocyst knobs. With 4 radial canals, and 8 tentacle bulbs. Tentacle numbers alternate on bulbs, with 3–5 tentacles on each bulb at the base of a radial canal and 1–3 tentacles on the alternate bulbs. Tentacle bulbs and stomach yellowish to reddish to brown; without ocelli. *Natural History*: Prey include young copepods, cladocerans, rotifers, and tunicate larvae. *Range and Habitat*: North-boreal species in Atlantic and Pacific Oceans; Bering Sea to Mission Bay.

Family Cladonematidae

13. *Cladonema californicum*

Identification: Bell up to 2–3 mm in diameter and rounded; mouth with 6 short oral arms with terminal nematocyst knobs. With 6–7 gonads emerging from about the midpoint of the manubrium as elongated, rounded protrusions in a whorl. Usually with 9 unbranched radial canals, and the same number of marginal tentacles. Each marginal tentacle has 2 or 3 branches. Velum thick and muscular. With a reddish ocellus at the base of each tentacle. *Natural History*: Occurs on mud flats covered with *Ulva* that are more or less exposed at very low tide, spending most of its time attached to seaweed. Feeds on small crustaceans. *Range and Habitat*: Found in protected bays and inlets from southern California to British Columbia. *Remarks*: The European species *C. radiatum* has become established in some bays.

Family Corynidae

14. *Sarsia* spp.

Identification: Bell to 6–20 mm tall (occasionally to 40 mm); with 4 narrow radial canals and 4 tentacles each with a prominent ocellus. Manubrium long and slender. Gonad completely encircling manubrium for most of its length. Mostly colorless, but often tinged with blue, red, green, or orange on manubrium, tentacles and/or tentacle bulbs. *Natural History*: Often abundant in nearshore waters. Usually found spring through fall, but in most regions more than one species occur, often sequentially. They fish by hanging passively with their tentacles extended and feed especially on crustacean zooplankton. The hydroids form robust, easily seen colonies in the spring that are often pink in color, on floats or rocky substrates. *Sarsia* can be very difficult to identify to species. *Range and Habitat*: Common in coastal surface waters from central California to the Bering Sea.

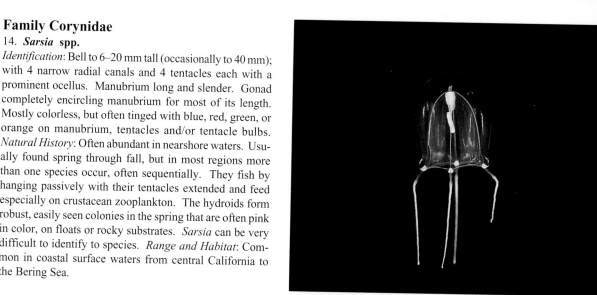

Family Moerisiidae

15. *Maeotias inexspectata*

Identification: Up to 56 mm in diameter, usually less than 40 mm, with manubrium large, with 4 very long frilly lips; no gastric peduncle. With 4 radial canals, each with curtainlike gonads that extend onto "arms" of the manubrium out along each radial canal. Ring canal has several prominent centripetal canals extending upward in each quadrant. With up to 600 tentacles. With numerous statocysts, each with an indistinct ocellus, around the bell margin. Umbrella is milky white, with a pink blush around the bell margin. *Natural History*: Medusae are released in late spring and early summer and then grow up together as a cohort. The medusae usually rest on the bottom, only sometimes rising to the surface. Feeds on small crustaceans. *Range and Habitat*: Originally from the Black Sea; found on the west coast in low salinity sloughs off San Francisco Bay.

29

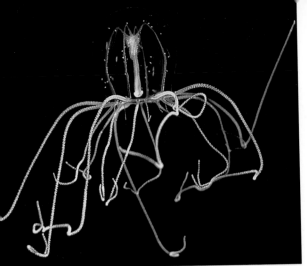

Family Polyorchidae

16. *Polyorchis haplus*

Identification: Up to 20 mm high, with conical gastric peduncle protruding into the bell cavity, from which long tubular gonads hang, up to 25 on each radial canal; stomach long, with 4 frilly lips. The 4 radial canals are simple in most specimens and branched in only the largest ones, with rudimentary, closely-set, knob-like diverticula; the ring canal with or without knob-like diverticula. With up to 30 tentacles. With deep red pigment around the ocellus at the base of each tentacle; tentacles and gonads typically tinged yellow or grayish brown. *Natural History*: Sympatric with, but usually less common than, *P. penicillatus*. Found year round, but relatively rare; polyp not known. *Range and Habitat*: Santa Monica Bay to Bodega Bay, central California.

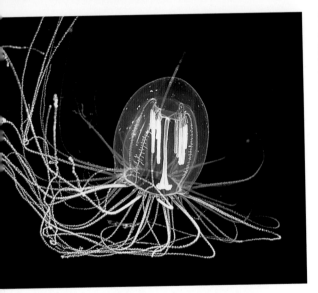

17. *Polyorchis penicillatus*

Identification: Up to 40–60 mm high with rounded gastric peduncle protruding into the bell cavity, from which long tubular gonads (≤15) hang on each radial canal; stomach long, with 4 frilly lips. The 4 radial canals are branched, each with 15–25 pairs of lateral diverticula, the ring canal may also have a few, knob-like centripetal diverticula. Up to 160 tentacles, usually less than 100, closely packed; without marginal vesicles. With dark ocelli surrounded by red pigment at the bases of the tentacles; umbrella, manubrium, gonads and tentacles usually whitish, but may be purplish, orange or yellow. *Natural History*: Occurs in bays, where the medusae spend most of their time on the bottom. Feeds mostly on benthic crustaceans and demersal zooplankton. Older medusae sometimes have unicellular green algae growing on the exumbrellar surface. The polyp remains unknown. *Range and Habitat*: Found along the entire Pacific coast of North America, from the Sea of Cortez and San Diego to the Aleutian Islands.

18. *Scrippsia pacifica*

Identification: Up to 100 mm high with conspicuous gastric peduncle more than half the length of the bell cavity, and with 30–60 long tubular gonads suspended from the peduncle on each radial canal. Stomach short, with 4 frilly lips. The 4 radial canals are mostly unbranched. With up to 256 tentacles, in 7 cycles, the largest projecting from some distance up the sides of the exumbrella. With ocelli at the bases of the smaller tentacles; without marginal vesicles. Umbrella transparent, manubrium, gonads and tentacles whitish to pale yellow; canals white, yellow, or carmine. *Natural History*: Lives deep offshore in bays. Moribund specimens are found at the surface or on nearby beaches in spring and early summer. *Range and Habitat*: Bahía de Sebastián Vizcaíno, Baja California to northern California.

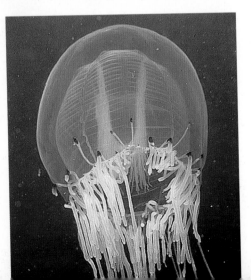

Family Euphysidae

19. *Euphysa* spp.

Identification: Bell to about 12 mm tall; with 4 radial canals. With 4 rounded tentacle bulbs and with 1–4 tentacles (which may be different lengths); without ocelli. Manubrium tubular; gonad completely encircles manubrium. Bell transparent; manubrium, tentacle bulbs, and tentacles often with scarlet pigment, but may be white, yellow or orangish. *Natural History*: Seen nearly any month of the year, but uncommon in most localities. Two to four different species may be present along the west coast, some characteristic of shallow water and others of deep water. Most probably have a solitary polyp that lives in mud or interstitially in fine gravel. *Range and Habitat*: California to the Bering Sea. *Remarks*: Species found on the west coast include *E. flammea* and/or *E. japonica* (which have 4 equal tentacles) and *E. tentaculata* (only 3 tentacles).

R. Miller

Family Velellidae

20. *Porpita porpita*   Blue button

Identification: Floating hydroid colony covered by a chitinous float (usually only a few cm diameter) and surrounded by a membrane that lies on the water surface film. Float is usually golden brown and mantle and hydroid zooids beneath it vary from deep turquoise blue to lemon colored. Medusae, liberated from the hydroid stage, are 0.3–2.5 mm tall. They have 8 radial canals with yellow-brown zooxanthellae and only 2 opposite tentacles, each terminating in a large nematocyst-covered ball. *Natural History*: *Porpita* hydroids are the best-known stage. They live on the water surface and feed on other small animals that live in that nearly two dimensional ecosystem. The medusae are rarely seen in the field. *Range and Habitat*: Cosmopolitan in warmer waters. Sometimes washed ashore on beaches of Baja California.

21. *Velella velella*   By-the-wind sailor

Identification: Floating hydroid colony covered by an elliptical to rectangular float (usually less than 6 cm long) with a triangular sail held above the water surface, surrounded by a membrane that lies on the surface film. Float and mantle are deep-blue colored. The hydroid colony beneath them consists of several types of zooids, all very brownish from the many zooxanthellae inside the tissue. Medusae, liberated from the hydroid stage in great numbers, are 1–3 mm tall and brown-colored from the zooxanthellae located in the subumbrella. They have 4 radial canals, but only 2 opposite tentacles. *Natural History*: The floating hydroids come ashore on beaches all along the west coast, usually in late spring or early summer. The hydroid eats fish eggs, euphausiid eggs, copepods, and appendicularians, and is eaten by many surface-dwelling animals. *Range and Habitat*: Cosmopolitan, usually in warm to warm-temperate waters.

Order Leptomedusae
Family Aequoreidae

22. *Aequorea* spp.

Identification: Bell diameter up to 25 cm; with many radial canals, and with gonads extending along the radial canals. With about 1/2 to 3 times as many tentacles as radial canals, all alike, and fine. Generally colorless; bioluminescent around the margin. *Natural History*: Alaskan specimens approach 25 cm in diameter and have very large numbers of radial canals. A smaller species, reaching about 8 cm diameter, is found from British Columbia to central California, and has been variously called *A. aequorea, A. flava, A. forskalea,* or *A. victoria*. A species off Baja California can divide by fission. *Aequorea* eat mostly soft-bodied prey, including other hydromedusae, ctenophores, polychaetes, and appendicularians. Hydroids only infrequently collected in the field. *Range and Habitat*: Entire west coast, from late spring through autumn.

Family Blackfordiidae

23. *Blackfordia virginica*

Identification: Up to 18 mm in diameter, usually less than 15 mm. Stomach small, with 4 long lips. With 4 radial canals, each bearing a gonad extending out from a corner of the stomach more than half the length of the radial canal. With up to 80 tentacles, each with a distinctive marginal bulb that has a small fingerlike endodermal projection extending inward from the bell margin into the gelatinous substance of the bell. With 1 or 2 statocysts between every two tentacles, without ocelli. Transparent with whitish gonads and sometimes with black pigment along bell margin. *Natural History*: Tolerates salinities 3–35‰ and is eurythermal. Feeds on copepods, copepod nauplii and barnacle nauplii. *Range and Habitat*: Probably native to the Black Sea, but now worldwide, mostly in bays and harbors where it has been introduced by shipping. Known on the west coast in low salinity portions of San Francisco Bay and Coos Bay, Oregon.

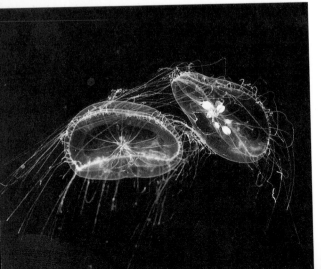

Family Dipleurosomatidae

24. *Dipleurosoma typicum*

Identification: Bell to 12 mm wide, flabby. Central stomach variable in shape, with small lips, and with 5–18 radial canals, irregular in distribution, may be branched, and sometimes not all reaching the margin. With 1–12 gonads, each on a radial canal near the stomach. With more than 100 tentacles in large individuals, each with a small black ocellus on the inside. Stomach and lips with blackish pigment, gonads white, tentacle bulbs brownish-red. *Natural History*: One of a small number of hydromedusae that reproduce asexually by fission of the bell, which explains the variable shape of the stomach and number of radial canals. Feeds primarily on copepods. *Range and Habitat*: North Atlantic and North Pacific, known from Friday Harbor to SE Alaska.

Family Melicertidae

25. *Melicertum octocostatum*

Identification: Bell to 14 mm high; stomach short and broad, with base attached to subumbrella and continuing down each radial canal, with 8 lips. With 8 radial canals, each with a sinuous linear gonad. With up to 88 larger tentacles alternating with about as many small ones. Without marginal vesicles and without ocelli. Umbrella transparent, manubrium and gonads yellow or brownish-yellow. *Natural History*: Coastal species, of cold, and sometimes estuarine, water. Relatively weak swimmer, sometimes hanging upside down; feeds on copepods, polychaete larvae, and fish larvae. *Range and Habitat*: Known principally from the North Atlantic, it also occurs widely in the North Pacific; Friday Harbor and British Columbia in late spring, Sitka, Alaska and Bering Sea in late summer, and central Oregon.

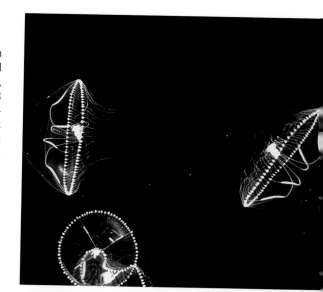

Family Eirenidae

26. *Eutonina indicans*

Identification: Up to 25–35 mm bell diameter; manubrium with 4 frilly lips hangs below the bell margin on a conical peduncle. With 4 radial canals; gonads linear and sinuous, on radial canals. With about 200 short tentacles and only 8 marginal vesicles. Umbrella transparent; stomach and gonads bluish white or slightly tinged with yellow; tentacles with a little black pigment at bases. *Natural History*: Most abundant in the spring through autumn. Feeds on invertebrate eggs and larvae, appendicularians, and small hydromedusae. Small hydroid colonies (less than 200 polyps) have been found on *Zostera*, subtidal rock and a crab carapace. *Range and Habitat*: Pacific coast at least from Santa Barbara to Vancouver Island, Aleutian Islands, Bering Sea, Kamchatka, and Hokkaido. Usually occuring in great numbers near the surface nearshore and on the shelf.

Family Laodiceidae

27. *Ptychogena* spp.

Identification: Bell diameter from less than 10 mm up to 90 mm, from nearly hemispherical to flattened. With 4 radial canals that give rise to the many lateral diverticula that contain the gonads. Stomach square, opening wide with 4 simple lips and with 4 broad funnel-shaped lobes leading into the radial canals. With up to 500 tentacles. With club-shaped marginal vesicles between the tentacles. *Range and Habitat*: Two different species, *Ptychogena californica* and *P. lactea*, might be collected from San Diego to the Bering Sea. They seem to be mostly in deep water off California, getting progressively shallower further north.

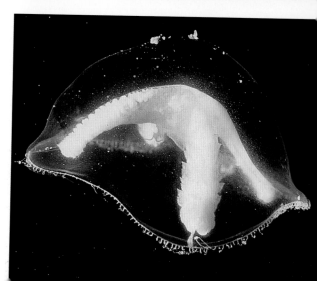

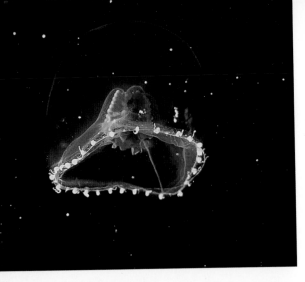

Family Tiarannidae

28. *Modeeria rotunda* (= *Tiaranna rotunda*)

Identification: Up to about 30 mm in bell diameter; stomach in mature specimens becomes highly arched and extends above the general subumbrellar cavity; mouth with 4 large ruffled lips. Gonads in regular folds on the walls of the stomach and attached to the 4 radial canals. With up to 40 tentacles, and with 1–3 spindle-shaped structures between each tentacle. Bell transparent and colorless; stomach, lips and gonads deep brick-red. *Natural History*: The thecate hydroid of this medusa has been collected living on other hydroids in water 35–500 meters deep. *Range and Habitat*: Uncommon deep-water species found in the north Atlantic, Mediterranean, Pacific and Antarctic Oceans; it has been collected in the San Clemente Basin off San Diego. *Remarks*: Might be confused with *Chromatonema rubrum*.

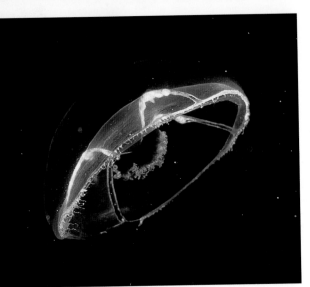

Family Mitrocomidae

29. *Foersteria purpurea*

Identification: Up to 30–40 mm in diameter, stomach attached to a low peduncle, with 4 long and very frilly lips. With 4 radial canals, each bearing much folded, curtain-like gonads. With up to about 150 tentacles, all alike, in large specimens. With about 100 marginal vesicles, without ocelli. Umbrella transparent, gonads and radial canals white or purple, manubrium and lips dark purple. *Natural History*: Observations of this species from a submersible in British Columbia (C.M.) revealed the medusae hovering within a few meters of the bottom. Most were motionless, and those that swam pulsated slowly. Specimens varying from newly released (a few mm in diameter with only 4 tentacles) to mature adults were all present at the same time. *Range and Habitat*: Near bottom in British Columbia inland marine passages, mostly 200 meters and deeper; Monterey Bay.

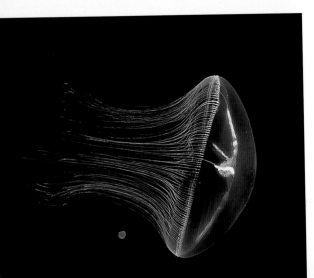

30. *Mitrocoma cellularia* (= *Halistaura cellularia*)

Identification: Up to 90 mm in diameter; with an irregularly creased exumbrellar surface. Stomach small with 4 long frilly lips. With 4 radial canals, each with gonads along nearly their entire lengths, but often not developing until the growing medusa is well over 20–30 mm diameter. With up to 340 tentacles, the tentacles all alike in mature specimens, but with 1–3 marginal cirri between tentacles in small animals. With 16–24 marginal vesicles; without ocelli. Umbrella slightly opaque, gonads white. *Natural History*: At Friday Harbor, medusae are found from early spring to mid-autumn. In Monterey Bay, they are most common fall through early spring. The medusae are strongly bioluminescent, localized in a narrow band around the margin. *Range and Habitat*: Point Barrow, Alaska and Bering Sea to central California.

31. *Mitrocomella polydiademata*

Identification: Up to 30 mm in diameter, but usually less than 20 mm; stomach small, mouth with 4 short, simple lips. With 4 narrow radial canals, each bearing linear gonads. With 36–64 tentacles and with groups of 5–9 or more cirri between every pair of tentacles. With 12–16 marginal vesicles without ocelli. Bell transparent with pale pink gonads, radial canals, bulbs, tentacles, and stomach. *Natural History*: Short-lived species of late spring, surviving approximately one month. Frequently found in the guts of *Aequorea victoria*; the very long tentacles may make it particularly susceptible to capture. Luminescent, but lacks the green fluorescent protein common to most hydromedusae, so it luminesces blue rather than green. The colonial hydroid has been raised in the laboratory. *Range and Habitat*: Common in Friday Harbor and southern British Columbia in May. Also North Atlantic and Arctic Ocean.

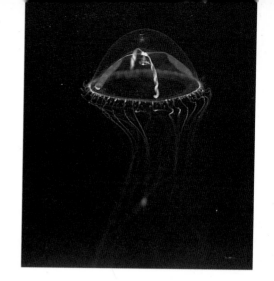

Family Tiaropsidae

32. *Tiaropsidium kelseyi*

Identification: Up to 80 mm in diameter, usually less than 50 mm; with short quadrate stomach and somewhat frilly mouth; no gastric peduncle. With 4 radial canals, each bearing folded, curtain-like gonads along nearly the entire length. With up to 16 tentacles, each having an elongate marginal bulb, and up to 100 rudimentary tentacles. With 8 marginal vesicles, each with a black ocellus. Umbrella very transparent; manubrium, gonads and tentacle bulbs with some pale yellowish or dark gray pigment. *Natural History*: Since it is so infrequently collected, this species may be oceanic rather than coastal; some specimens have been taken from deep open trawls. Hydroid unknown. *Range and Habitat*: Known only from the west coast of North America. Collected occasionally from San Diego to southern Vancouver Island.

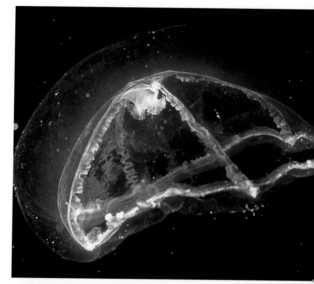

R. Miller

33. *Tiaropsis multicirrata*

Identification: Up to about 30 mm in diameter, but usually not more than 15–18 mm. With small stomach on a broad low peduncle; mouth with 4 long, very frilly lips. With 4 radial canals, each bearing straight or sinuous gonads along middle 1/2–4/5 of the length. With about 300 short, fine tentacles; with 8 marginal vesicles, each with a black ocellus. Umbrella transparent; stomach, gonads and tentacle bases dull yellow tinged with black. *Natural History*: Sometimes occurs in mass aggregations; feeds on other hydromedusae as well as on crustacean zooplankton and planktonic larvae. The hydroid grows on the stems of other hydroids or on algae. The eggs of this species are said to develop to the planula stage within the gonads of the medusa. *Range and Habitat*: North boreal Atlantic and Pacific. On the Pacific coast from the Aleutian Islands to southern Vancouver Island in the spring and summer.

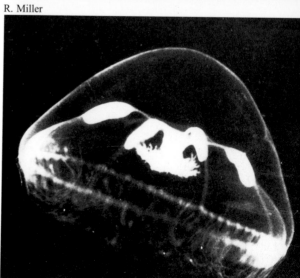

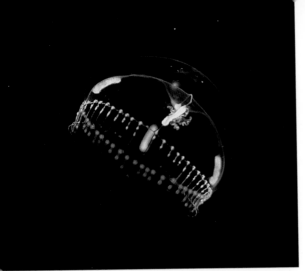

Family Campanulariidae

34. *Clytia gregaria* (= *Phialidium gregarium*)

Identification: Up to 20 mm in bell diameter; stomach small, with 4 long ruffled lips. With 4 radial canals, each bearing a linear gonad along its distal half to two-thirds. With up to 65 tentacles; with 1–2 marginal vesicles between each pair of tentacles. Umbrella transparent, gonads, tentacle bulbs, and manubrium with some pale yellowish or brown pigment, and frequently with a black pigment ring around the margin. *Natural History*: One of the most abundant small hydromedusae in the Pacific Northwest. Its hydroid is an inconspicuous part of the fouling fauna. Medusae are released over a period of many months from spring until early fall, and frequently host the parasitic sea anemone larva *Peachia quinquecapitata*. *Range and Habitat*: Coastal, from central Oregon to the Bering Sea, often in great numbers.

35. *Obelia* spp.

Identification: Medusa flat, to 6 mm but usually less, with overall appearance of an inside-out umbrella. Small central stomach with 4 simple lips; 4 radial canals, each with a round sac-like gonad. With more than 16 short stiff tentacles; with 8 statocysts. *Natural History*: *Obelia* hydroids are common members of the float-fouling fauna. The tiny medusae are most likely to be encountered in plankton tows. Although frequently collected, little is known about their natural history. *Range and Habitat*: Genus has a worldwide distribution; medusae can be found nearshore along the entire west coast from mid-spring through mid-autumn. *Remarks*: There are several special of *Obelia* on the west coast. The taxonomy of the hydroids is still in some turmoil, and the medusae change morphology as they grow, making specific identification even more difficult.

Order Limnomedusae
Family Olindiasidae

36. *Aglauropsis aeora*

Identification: Up to 20 mm in diameter; manubrium has 4 frilly lips bordered by a row of closely-packed nematocyst batteries. With 4 broad radial canals and folded curtain-like gonads extending the entire length of each canal; ring canal broad; no centripetal canals. With about 200 tentacles of varying sizes, emerging via grooves in the jelly at various heights above the bell margin, ringed with nematocysts. With 60–100 statocysts. Transparent with pale blue-green gonads and manubrium, pink-brown lips and tentacles. *Natural History*: Fairly mature specimens have been collected only in the summer and autumn in bays and washed up on beaches. *Range and Habitat*: Known only from Bodega Bay, Tomales Bay, and Monterey Bay in central California; may be from slope water or oceanic.

37. *Eperetmus typus*

Identification: Up to 45 mm wide; long manubrium; mouth with 4 short frilly lips lined with nematocyst knobs; very low gastric peduncle. With 4 broad radial canals each bearing a wavy, curtainlike gonad. Ring canal with up to 6 broad centripetal canals in each quadrant. About 100 tentacles, emerging from grooves in the jelly at various heights above the bell margin, ringed with nematocysts. With numerous marginal statocysts, alternating with the tentacles. Umbrella transparent, manubrium, gonads and tentacles delicate pink. *Natural History*: Collected infrequently in coastal waters, spring through early winter; may be oceanic or slope water. The medusae have a strong "crumple" response to being disturbed, withdrawing all tentacles up under the contracted bell. Polyp not known. *Range and Habitat*: Known from southeast Alaska, Vancouver Island, northwest Washington, and Coos Bay and Yaquina Bay in Oregon. Uncommon.

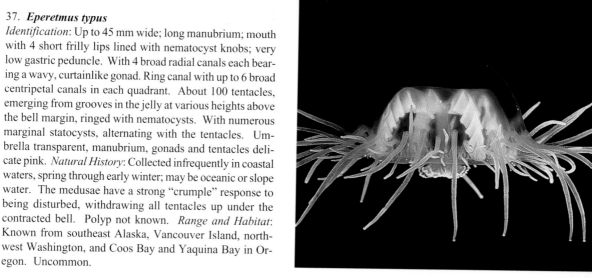

38. *Gonionemus vertens*

Identification: Bell diameter to about 25 mm, stomach with 4 short, frilly lips. With 4 radial canals, each bearing a gonad along most of its length; without centripetal canals. With up to 80 tentacles, each with an adhesive disk near the middle, and with rings of nematocysts, and about as many statocysts as tentacles. Bell transparent, with bright orange to yellowish-tan gonads and tentacle bulbs; pale yellow manubrium. *Natural History*: Medusae called *G. vertens* exist on both sides of the north Pacific. Those in the Russian Far-East are extremely venomous, whereas those from the Pacific Northwest do not sting. West coast medusae are found in the summer living amongst and frequently attached to eelgrass, *Ulva*, or laminarian algae; they feed on copepods. *Range and Habitat*: Protected waters in Washington, British Columbia, Aleutian Islands, northern Japan and Kamchatka.

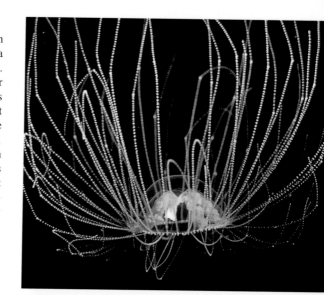

39. *Vallentinia adherens*

Identification: Bell diameter about 8 mm; stomach short and quadrangular. With 4 radial canals, each bearing a gonad shaped like a ruffle, no centripetal canals. With 4 long tentacles having terminal adhesive disks and 40 or more shorter tentacles with rings of nematocysts and usually also with an adhesive pad near the distal end, with small cirrus-like tentacles with rings of nematocysts alternating with the other tentacles. With 1 or 2 statocysts between successive tentacles. Bell transparent with creamy white manubrium, golden brown radial canals, and beige tentacles. *Natural History*: Found clinging to algae in shallow water, nearshore. *Range and Habitat*: Known only from Pacific Grove, California.

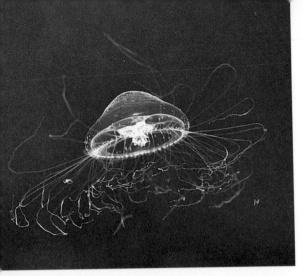

Family Proboscidactylidae

40. *Proboscidactyla flavicirrata*

Identification: Bell to 12 mm wide and 10 mm high; stomach covered with 4 gonads that extend a short distance out the radial canals, with highly folded lips. With 4 radial canals that branch irregularly resulting in 50–70 terminal branches reaching the bell margin, and with the same number of tentacles or a few more. With up to 60 small clusters of nematocysts on the lower part of the exumbrella. Tentacle bulbs usually golden-brown in life, gonads creamy white to tan. *Natural History*: Feed especially on small gastropod and bivalve larvae. Hydroid is an obligate commensal on curled tip of the tube of several species of sabellid worms. *Range and Habitat*: Coastal North Pacific from Oregon to China. Common in Puget Sound and SE Alaska throughout the summer.

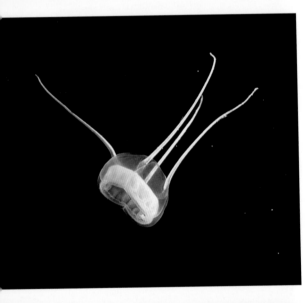

Order Narcomedusae
Family Aeginidae

41. *Aegina citrea*

Identification: Bell hemispherical to conical and to about 50 mm in diameter (usually less than 20 mm). Typically with 8 rectangular stomach pouches. With 4 (sometimes 5 or 6) tentacles that have a very evident, curved "rooted" portion within the jelly. With 4 (may be 5 or 6) marginal lappets, each with several marginal statocysts. Bell transparent and clear, or may be suffused with lemon yellow color. Some specimens have a reddish tentacle epithelium, which is easily damaged. *Natural History*: Preys on gelatinous zooplankton such as other hydromedusae, ctenophores, salps and appendicularians. It is an active swimmer and often holds the 4 tentacles high above the bell. *Range and Habitat*: Widely distributed in warm and temperate parts of the ocean. *Remarks*: Similar medusa with only 2 tentacles is *Solmundella bitentaculata*.

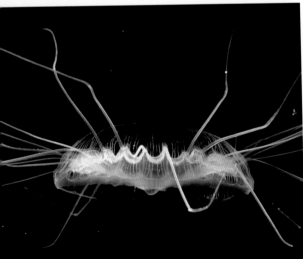

Family Solmarisidae

42. *Pegantha* spp.

Identification: Bell rounded to fairly flat, with thick, lens-like jelly; to 30–50 mm in diameter. Without stomach pouches, but the gonadal portion of the circular stomach may bulge into the lappets, in some cases giving a crenulated pouch-like appearance. With 12-40 tentacles, depending on the species and age, and the same number of marginal lappets. With several statocysts on the margin of each lappet and with several linear extensions or tracts going from the margin some distance up the exumbrella on each lappet. Exumbrella sometimes with ridges. Transparent and colorless, or may be suffused with yellow, violet-pink, or blue color. *Range and Habitat:* Several different species can be found in all the oceans between 40°N and 40°S. Some are epipelagic, whereas others occur in deep water.

43. *Solmaris* spp.

Identification: Bell flat and lens-shaped, less than one cm in diameter. Without stomach pouches; stomach has an unbroken circular peripheral boundary; gonads develop in the walls of the stomach. With 10–36 tentacles, exiting the bell above the margin opposite the periphery of the stomach, each with a short "rooted" portion within the jelly. With as many marginal lappets as tentacles, each with about 1–3 statocysts. Usually transparent and colorless. *Natural History*: Often seen in great numbers near the surface off central and southern California. They are active swimmers, usually swimming in bursts of pulsations. Prey includes other small gelatinous zooplankton including chaetognaths and doliolids. *Range and Habitat*: Shelf or oceanic epipelagic, usually in warmer waters. A number of species of *Solmaris* occur worldwide; all are very small.

R. Gilmer

Family Cuninidae

44. *Cunina* spp.

Identification: Bell to about 60 mm in diameter with 6–29 stomach pouches and the same number of tentacles and lappets. Tentacles leave upper surface of the exumbrella opposite the center edge of each stomach pouch. With 2–5 evenly-spaced statocysts at the margin of each lappet, each with a small vertical extension on the bell. Transparent and colorless, or suffused with pink, magenta, or purple color. *Natural History*: Occasionally seen in surface waters along the west coast at least as far north as Washington. May carry developing larvae of *other* Narcomedusae in its stomach pouches. Prey include other gelatinous zooplankton such as other hydromedusae, ctenophores, salps and doliolids. *Range and Habitat*: Oceanic, mostly epipelagic, in warmer waters; relatively uncommon and poorly known.

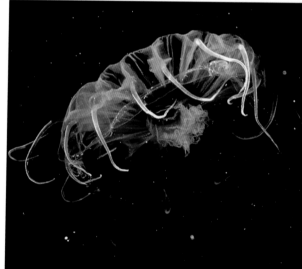

45. *Solmissus* spp.

Identification: Bell up to about 100 mm in diameter, the gelatinous disk is quite thick. With 8–40 stomach pouches that are oval to rectangular in outline; mouth is a simple round opening in the center. With as many lappets and tentacles as stomach pouches, lappets square to rectangular, each with up to 2–15 evenly spaced marginal statocysts. Tentacles have a well-defined conical insertion base. Usually colorless and transparent, but may be suffused with magenta or other pinkish or bluish hue. *Natural History*: May be either swimming actively or quiescent in the water column. Tentacles usually curve up and out from the top of the bell. Feeds primarily on gelatinous zooplankton including medusae, ctenophores and salps. *Range and Habitat*: Worldwide in warm and temperate seas, from the surface to abyssal depths.

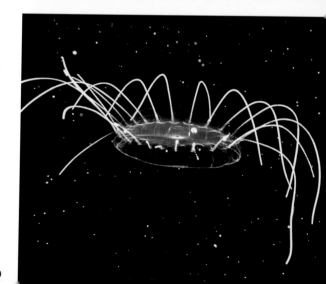

Order Trachymedusae
Family Geryoniidae

46. *Geryonia proboscidalis*

Identification: Bell to 80 mm in diameter, usually much less; stomach small, with 6 simple lips, attached to a long conical gelatinous peduncle. With 6 radial canals that continue down the peduncle to the stomach. Six flat heart-shaped gonads on the radial canals, each separated by up to 7 small canals. With 6 long tentacles alternating with 6 small tentacles; with 12 statocysts. Transparent and colorless. *Natural History*: The long peduncle and manubrium is very active, and rapidly responds to touch on the edge of the bell by swinging over and bringing the mouth directly into contact with the area touched. *Geryonia* is usually less common than *Liriope*. *Range and Habitat*: Upper layers of all tropical and subtropical oceans, occasionally as far north as central California.

47. *Liriope tetraphylla*

Identification: Bell to 30 mm in diameter; stomach small, with 4 simple lips, attached to a long conical gelatinous peduncle. With 4 radial canals that continue down peduncle to stomach. Four flat leaf-shaped gonads on the radial canals, each separated by up to 3 broad canals that come up from the margin. With 4 long tentacles alternating with 4 small tentacles; with 8 statocysts. Transparent and colorless. *Natural History*: Although reported frequently in many parts of the world, little is known about its biology. Occurs in great numbers at the surface when warm oceanic water comes nearshore in central and southern California. Its nematocysts can be mildly irritating, causing a form of dermatitis. *Range and Habitat*: Distributed from about 40°N to 40°S. On the west coast, occurs in surface waters north to about Bodega Bay, California, increasing in frequency to the south; common in the Sea of Cortez.

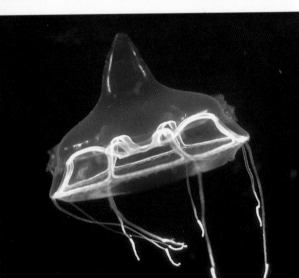

Family Halicreatidae

48. *Halicreas minimum*

Identification: Bell to 40 mm wide and not as high, with pointed apical projection of varying size and with 8 clusters of gelatinous projections on the exumbrella overlying the radial canals; jelly very stiff. Circular mouth and stomach; 8 broad radial canals that are swollen by the long, flat gonads; broad velum. Up to 640 tentacles, the largest at the perradii, with nematocysts concentrated on the distal portion, and about as long as the bell diameter; with up to 8 pendant statocysts per octant. Transparent and colorless, or with pale orange color in the canals and stomach. *Natural History*: A weak swimmer whose stiff jelly allows only the lowermost part of the bell to deform during swimming contractions. Canals are sometimes lined with small oil droplets. *Range and Habitat*: Deep water, cosmopolitan.

49. *Haliscera* spp.

Identification: Bell to 24 mm wide with rounded to conical apical projection. Circular mouth and stomach at the intersection of 8 broad radial canals that are swollen by flat, oval gonads in their midportions; broad velum. With up to 160 tentacles that have nematocysts concentrated on the distal portion and are 1–3 times as long as the bell diameter; with 2–5 pendant statocysts per octant. Transparent and colorless, or with orange color in the canals and gonads and sometimes with a rose-pink stomach and mouth. *Range and Habitat*: Deep water in both the Atlantic and Pacific Oceans. *Remarks*: Both *H. bigelowi* and *H. conica* are commonly collected in mid-water trawls off the west coast.

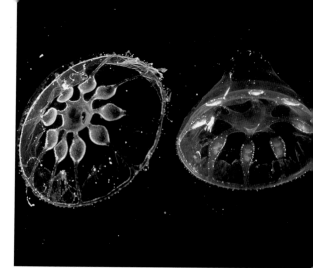

H. bigelowi   *H. conica*
R. Gilmer

50. *Halitrephes maasi*

Identification: Bell to about 100 mm wide. Circular mouth and stomach; more than 16 broad radial canals, some of which branch dichotomously, with total canal numbers sometimes more than 30. Gonads along radial canals. With up to 70 tentacles of different sizes, with nematocysts concentrated on the stiff distal portion. The endodermal roots of most tentacles can be counted in even the most net-battered specimens. With up to 80 or more pendant statocysts. Transparent and colorless, or with pale orange canals, stomach and tentacles. *Natural History*: This large hydromedusa is very active, capable of long swimming bouts of strong contractions. Canals are sometimes lined with small oil droplets, indicating recent feeding on lipid-rich, probably crustacean prey. *Range and Habitat*: Deep water, probably nearly cosmopolitan in warm and temperate regions. *Remarks*: The less-common *H. valdiviae*, with 15-16 radial canals and up to 200 tentacles, some of which have a large terminal ball, also occurs on the west coast.

Family Ptychogastriidae

51. *Ptychogastria polaris*

Identification: Bell to 22 mm wide, with 16 radiating ridges. Stomach with 8 lobes attached to the 8 radial canals by mesenteries; with 16 gonads (one on either side of each gastric lobe); mouth with 4 simple lips. Velum very wide and muscular. With about 48 clusters of tentacles in adult, each cluster with both filiform and adhesive tentacles; with 16 marginal statocysts. With yellow gonads and a red manubrium. *Natural History*: Typically sits on the bottom using its inside adhesive tentacles for attachment, and with its filiform tentacles extended outward in a static fishing posture. *Range and Habitat*: Circumpolar Arctic and Antarctic as shallow as 10 meters; known from deep water in some British Columbia fjords and Monterey Bay.

G. Dietzmann

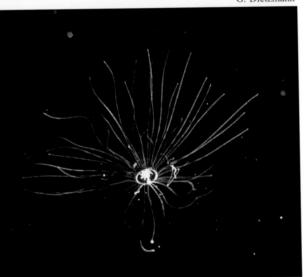

Family Rhopalonematidae

52. *Aglantha digitale*

Identification: Bell to 40 mm high, usually less than 20; stomach with 4 simple lips attached to a long slender gelatinous peduncle, with 8 radial canals that continue down peduncle to stomach. Eight pendant gonads attach to the radial canals near the top of the peduncle. With 80 or more tentacles, with 8 club-shaped marginal statocysts. Transparent and colorless, or may have tinges of orange, pink or red. *Natural History*: Although usually epipelagic, *Aglantha* is a vertical migrator, sometimes penetrating into the upper mesopelagic zone by day and returning up near the surface to feed at night. It eats mostly copepods. Has 2 modes of swimming: a fast, powerful escape swim and a slow-swimming mode that produces a sort of cyclical swim-rest-swim-rest fishing pattern. *Range and Habitat*: Arctic and subarctic waters; common to central Oregon in the upper 200 meters.

53. *Aglaura hemistoma*

Identification: Bell to 6 mm high and 4 mm wide, jelly very thin. Stomach with 4 simple lips attached to a slender gelatinous peduncle that hangs about 1/2 into the subumbrellar cavity. With 8 radial canals that continue down onto peduncle to stomach. Eight sausage-shaped gonads attached to the radial canals at about the point where the peduncle and stomach join. With up to 85 tentacles a little longer than the bell height, and with 8 club-shaped marginal statocysts. Transparent and colorless. *Natural History: Aglaura* replaces *Aglantha* as the most common epipelagic trachymedusa as one travels south. It has both an escape swim and a slower swim. Fragile and easily damaged, the tentacles break off near their bases from rough handling. *Range and Habitat:* Worldwide, in warm and warm-temperate near-surface waters from about 40°N to 40°S.

54. *Benthocodon pedunculata*

Identification: Bell to 40 mm wide with numerous fine furrows radiating toward the apex, evident on the lower portion of the exumbrella. Stomach with 4 flared lips. With 8 radial canals, each with a gonad running along and then suspended from about the midpoint, and with mesenteries connecting the gonads to the manubrium. With 1000–2000 crowded tentacles, the largest displaced upwards on the bell. With up to 30 marginal sensory clubs per octant. Exumbrella transparent with a little rusty pigment; subumbrella, manubrium, mesenteries and tentacles deep brownish-red; gonads white. *Natural History*: A strong swimmer. Eats small crustacea and foraminifera. *Range and Habitat*: Monterey Bay, San Clemente Basin, near mouth of Columbia River, Bahamas, Virgin Islands, in deep water, on or near the bottom most of the time.

42

55. *Colobonema sericeum*

Identification: Bell to 45 mm wide with conical subumbrella; velum broad. Stomach with 4 large lips; no gastric peduncle. With 8 radial canals and 8 gonads attached to the canals for most of their length. Apical outlines of the subumbrellar muscles form a star-shaped figure. With 32 tentacles and 32 statocysts, alternating. Without pigment; transparent. *Natural History*: This species is common in many regions. It is a very strong swimmer; the tentacles readily dehisce when it is disturbed – observers in submersibles have reported that released tentacles bioluminesce while the rest of the animal swims away. Frequently collected in mid-water trawls, but the tentacles are usually gone. *Range and Habitat*: Upper mesopelagic in the Pacific, Atlantic and Indian Oceans. *Remarks*: A smaller species with about 40 tentacles, *C. typicum*, may also be collected in deep water.

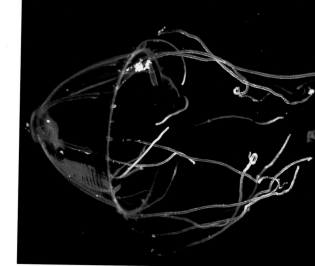

56. *Crossota rufobrunnea*

Identification: Bell to 15 mm wide; jelly soft; velum broad. Large manubrium with 4 recurved lips; no peduncle. With 8 radial canals, and 8 sausage-shaped gonads suspended from the canals near bell apex. With up to 250 crowded tentacles, all alike. Exumbrella transparent, with over 200 fine furrows running from margin to apex; subumbrella and manubrium dark brick red, tentacles brownish-red. *Range and Habitat*: Below 500 meters, Monterey Canyon, San Clemente Basin, off British Columbia; found in many locations in the North Atlantic and North Pacific Oceans. *Remarks*: *Crossota alba* is also found in deep water. It is distinguished by its transparent, unpigmented bell, nearly black manubrium, and 8 white to tan scimitar-shaped gonads suspended from below the midpoint of the canals. *Vampyrocrossota childressi* is about the same size, but black.

R. Harbison

57. *Pantachogon* spp.

Identification: Bell to about 12 mm high and wide, sometimes much less. With conspicuous musculature–the apical outlines of the subumbrellar muscle fields form a complete circle; broad muscular velum. Jelly thin, often with a small apical thickening. Manubrium tubular; mouth with 4 simple lips; peduncle. With 8 narrow radial canals, with gonads running nearly the entire length. With 48 or more tentacles, all alike, and with up to 8 statocysts per octant. Colorless and transparent or with varying amounts of orange color in the manubrium, gonads, velum, and/or subumbrella. *Range and Habitat*: Deep water, especially below 1000 meters, Baja California to British Columbia, and below 100 meters in some British Columbia fjords. *Remarks*: Several different medusae from deep tows off the west coast key out to *Pantachogon*. Some might be referable to *P. haeckeli*, others seem to be new species.

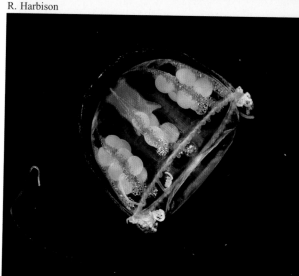

58. *Tetrorchis erythrogaster*

Identification: Bell to about 12 mm wide and often not quite as tall, pyriform, with thick apical jelly. Manubrium hangs anywhere from about 1/2 bell cavity to just below the bell margin, tubular; mouth with 4 simple lips; no gastric peduncle. With 8 narrow radial canals, and with pendant gonads attached at about the midpoint to only 4 of the canals. With 4 large perradial tentacles below the 4 gonad-bearing radial canals, and with 16–24 smaller tentacles. Statocysts apparently alternate with small tentacles. Velum broad and ruffled. Bell transparent and colorless, stomach deep carmine red, gonads white or tan, tips of large tentacles may be yellow. *Range and Habitat:* In deep and intermediate water in the tropical Pacific and Atlantic Oceans. Typically below 200 m off California; known from as far north as Monterey Canyon.

Subclass Siphonophorae

These exotic, chain-like carnivores are usually transparent and mostly oceanic, but landbound observers are in for a treat when these animals occasionally come inshore. A siphonophore may be as small as a few millimeters, or as long as many meters. Most have specific swimming behaviors that function to spread out their nematocyst-covered tentacles into a "net" for capturing prey. It is still debated whether siphonophores are colonies or individuals, but most specialists now prefer to think of them as individuals with many, well-integrated parts. Many of these parts are repeated multiple times by a budding process and are beautifully complex under the microscope; some are swimming bells, others serve for flotation, others are "stomachs," or have reproductive functions. These structures are mostly bilaterally symmetrical rather than radially symmetrical like the hydromedusae, to whom they are related.

Although there are only about 150 recognized species and many are cosmopolitan in distribution, the identification of siphonophores is not an easy task. There are relatively few images in the literature of entire animals, as most have been described from the various (mostly transparent) bits that become separated in plankton tows, either from the rigors of net collection, or from preservation in formaldehyde, which tends to separate even the nicest specimens into small pieces. Furthermore, many siphonophores bud off free-living sexual fragments, known as eudoxids, that swim away from their parent (or nurse) siphonophore, frequently even living at a different depth, and free-spawn gametes that develop into a new "nurse" siphonophore. All of the pieces of both the larger and eudoxid forms of each species must be described and connected with each other.

Because of the difficulties in identifying siphonophores to species, we have chosen to present only a few representative species in this guide, leaving the job of positive identification to the highly specialized taxonomic literature. For identification, we recommend first a small guidebook to British siphonophores by Kirkpatrick and Pugh (1984), which has considerable overlap with our west coast fauna; this can be augmented by the monographs of Totton (1965) or Bigelow (1911). The serious researcher will also need to consult the more recent, highly diverse, scientific literature. A review of siphonophore biology by Mackie *et al.* (1987) is particularly useful. Increasing publication of *in situ* siphonophore images taken by scientists using submersibles, ROVs, or even blue-water diving will help in the future for identification of intact specimens.

Order Cystonecta

Cystonect siphonophores are characterized by an apical gas-filled float, or pneumatophore, and they lack the swimming bells typical of other orders of siphonophores. Below the pneumatophore, cystonects have a stem region, or siphosome, which can be very long and is composed of a central stem, which buds multiple polypoid or medusoid structures. The float is either horizontal, as in *Physalia*, or vertical, as in the rest of the Cystonectae. Most if not all members of this small group have a sting that is painful to humans.

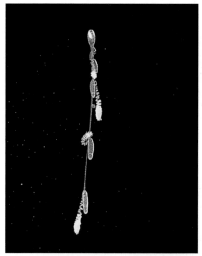

59. *Physalia physalis*   T. Heeger

60. *Rhizophysa* sp.   G. Dietzmann

Order Physonecta

Physonect siphonophores possess both an apical gas-filled float, or pneumatophore, and a close-fitting set of swimming bells, the nectophores. Below the swimming bells, physonects have a long stem region that includes both feeding and reproductive structures; these do not normally detach from the parent siphonophore, and their gametes mature in place. Most physonects are hermaphroditic, and some, including *Physophora*, can inflict painful stings. Some of the longest oceanic siphonophores are in this order.

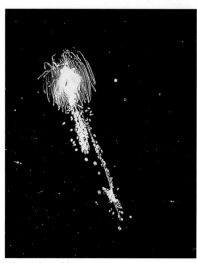

61. *Apolemia uvaria*

62. *Athorybia rosacea*   G. Dietzmann

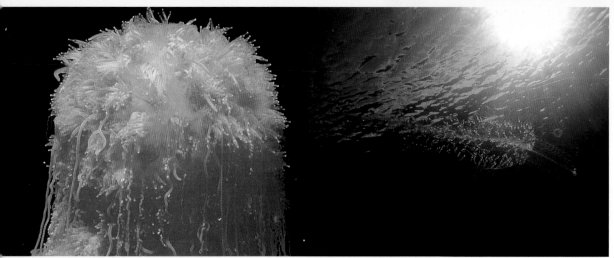

63. *Dromalia alexandri*   64. *Forskalia edwardsi*

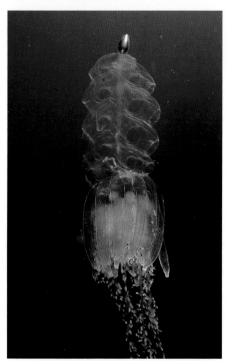

65. *Nanomia bijuga*   66. *Physophora hydrostatica*

Order Calycophora

This group lacks the apical gas-filled float, or pneumatophore, but has swimming bells, or nectophores. On the stem below the swimming bells, calycophorans have multiple identical units, each composed of several components, that can detach and become free-swimming sexual "eudoxids." Small calycophoran species may have a rocket ship-shaped swimming bell and can dart about quite rapidly; they have very precise behaviors for setting their tentacles as a fishing net (most are in the family *Diphyidae*).

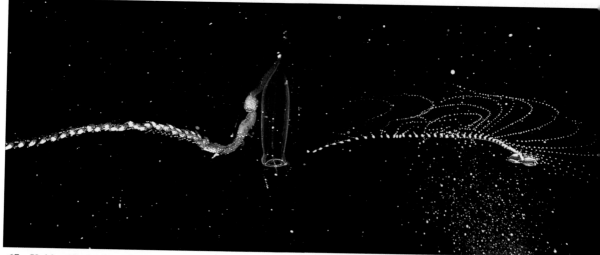

67. *Unidentified calycophoran*　　　　68. *Unidentified diphyid*

G. Dietzmann

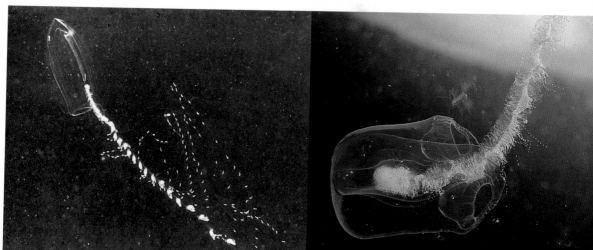

69. *Muggiaea atlantica*　　　　70a. *Praya sp. (swimming bells)*

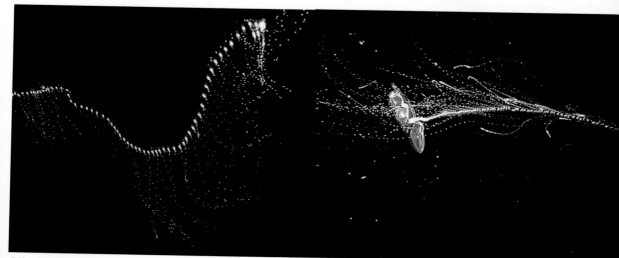

70b. *Praya sp. (stem)*　　　　71. *?Sulculeolaria sp.*

G. Dietzmann

Class Cubozoa

This primarily tropical group includes a number of species that superficially resemble hydromedusae. Because of the shape of the bell, these highly transparent medusae are also called box jellies. The so-called "sea wasps" are in this group and include the notorious Australian box jellyfish, *Chironex fleckeri*, whose sting can be fatal. Box jellies occur in a couple of discrete localities in southern California and otherwise are rare visitors within the range of this book; the best-known west coast representative lacks a potent sting.

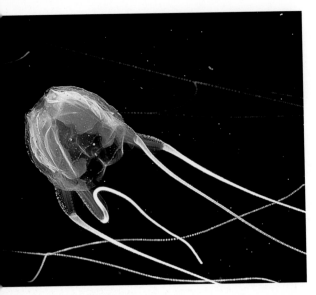

Family Carybdeidae

72. *Carybdea marsupialis*

Identification: Bell to 4 cm high; exumbrella with nematocyst-warts mainly in the interradii. Stomach small, with 4 simple lips. With 4 spatula-like pedalia arising at the bases of the 4 radial canals, each with a single heavy tentacle with rings of nematocysts. Clusters of gastric cirri at the 4 corners of the stomach, each arising from a single trunk (or rarely from 2). With light tan flecks on the bell. *Natural History*: Cubomedusae have well-developed eyes and are attracted to light. Swim almost continuously except when transferring prey to the mouth. Tentacles can extend 10 times the bell height. Prey include mysids, copepods and other crustaceans, polychaetes and small fishes. Swim near the bottom just inshore of kelps beds, from August to November. *Range and Habitat*: Tropical and subtropical; off Santa Barbara and La Jolla, possibly further south into Mexico; Atlantic and Mediterranean.

Class Scyphozoa

The scyphozoans include most of the larger types of jellies that people associate with the word "jellyfish". They can usually be distinguished from the more diminutive hydromedusae, particularly by their size and absence of a velum. With the exception of the sessile stauromedusae, most have free-swimming medusae. Scyphomedusae can be found throughout the world's oceans, but are never found in freshwater.

Scyphomedusae swim by contracting circular and radial muscles in the subumbrella of the bell. This expels water from the bell, pushing the animal forward. When the muscles relax, the elastic mesoglea (jelly) returns the bell to its resting shape. Coordination of bell pulsation is accomplished through the action of nerve cells of the subumbrellar network and the marginal nerve centers. Sensory information from the rhopalia along the bell margin can be relayed to the marginal nerve centers and influence the rate of pulsing. Each rhopalium is a complex sensory structure that may contain a statocyst, pigmented light-sensitive spots, and sensory pits lined with cells that may detect chemicals or scents.

Nematocyst-laden tentacles are generally used to capture a variety of zooplankton prey. Many scyphomedusae have potent stinging capability, although none of the species encountered along the west coast can cause serious harm to people. Oral arms are typically used to transport captured prey into the gastric cavity. Gastric filaments covered with nematocysts subdue any prey that remain alive after entering the gastric pouches, and also secrete digestive enzymes. Digestion of prey continues as the food is moved along the radial canals by cilia lining the entire gastric system. This system of canals distributes food and oxygen throughout the bell and also removes waste materials. Food is digested intracellularly following uptake by endodermal cells. In addition to being predators, scyphozoan jellies serve as food for a variety of animals, including other scyphomedusae, various fishes, sea birds and sea turtles.

Order Stauromedusae

Stauromedusae are quite unlike other scyphozoans in not having a free-swimming medusa stage. Even the planula larvae do not swim, but instead crawl until selecting a spot to attach. The planulae encyst and may spend several months dormant before emerging as a small polypoid form that does not strobilate like other scyphozoans, but develops directly into an adult "stalked jellyfish". Adults have a base much like a narrow anemone and live attached to various substrates such as seaweed and rock. Clusters of knobbed tentacles are used to capture small crustacean prey. Although stauromedusae are not capable of swimming, most have the ability to slowly move along the substrate and can also let go and drift to a new location.

Suborder Eleutherocarpida
Family Lucernariidae
73. *Haliclystus* spp.

Identification: Funnel-shaped, to nearly 3 cm wide. Stalk about the same length as calyx. Calyx with 8 equidistant arms, each with 30–100 capitate tentacles, alternating with 8 bean-shaped or trumpet-shaped marginal anchors on short stems. With 8 gonads, extending to the ends of the 8 arms; gonads with numerous round sacs, in regular or irregular rows. Subumbrella with or without folds. Color variable, in shades of green, brown, olive, yellow, orange, pink, red or purple, with white spots. *Natural History*: Found on a variety of intertidal and subtidal habitats, sometimes quite abundant, often matching the color of their substrate (typically rock, algae, or eelgrass). Feed on small epibenthic crustaceans. Late spring through autumn. *Range and Habitat*: North and South Atlantic; North Pacific from central California to Alaska, northern Japan and China; low intertidal and subtidal. *Remarks*: Larson (1990) assigned the name *H. octoradiatus* to the common *Haliclystus* along the west coast, whereas a Japanese specialist is planning to revive the name *H. sanjuanensis* for individuals in Washington and British Columbia. *H. salpinx* is clearly another species, of limited distribution in the San Juan Islands, and *H. stejnegeri* is a boreal species that occurs from Alaska to northern Japan.

R. Larson

Suborder Cleistocarpida
Family Depastridae
74. *Manania distincta*

Identification: Shaped like a goblet, with calyx clearly demarcated from the stalk. Calyx to 15 mm wide; stalk up to twice as long as the calyx. Eight short arms, in pairs, each with up to 26 capitate tentacles. With 8 small tentacles alternating with the arms. With 8 longitudinal paired gonads, each folded many times. Color light tan to cream, with four interradial dark-brown herringbone patterns on the calyx, extending as 4 dark lines down the stalk. *Natural History*: Found on seaweed, low intertidal and shallow subtidal. Feeds primarily on copepods. *Range and Habitat*: Northern Japan to Oregon, along the open coast in areas of wave action. Rarely collected in the eastern Pacific.

75. *Manania handi*

Identification: Shaped like a goblet, to 4 cm total length. Calyx longer than wide, or of nearly equal dimensions in life, and indistinctly demarcated from the stalk. Stalk highly contractile, stretching to approximately the length of the calyx when relaxed. Eight short arms, in pairs, each with 15–25 capitate tentacles alternating with 8 small tentacles. With 8 paired gonads, with obliquely-oriented folds. Color green to red with 4 interradial "windows" on the calyx of lighter color outlined in brown. *Natural History*: On *Zostera* and algae in semiprotected subtidal habitats from late summer through late spring; often with *Haliclystus*. The cryptic coloration makes this species difficult to find. Feeds on gammarid amphipods and harpacticoid copepods. *Range and Habitat*: Southern Vancouver Island, San Juan Islands, Puget Sound. *Remarks*: Another species, *Manania gwilliami*, also occurs on the west coast in exposed habitats. It is usually dark red, with bright white subumbrellar nematocyst vesicles, and can be very cryptic on coralline algae.

Order Coronatae

Coronate medusae are found near the surface in warm waters and at depth in colder waters (with some cold water species occurring near the surface at high latitudes). All have morphological similarities. The group is typified by an annular furrow, or coronal groove, in the exumbrella, located some distance down from the top, varying by species. This unique structure appears to act as a hinge, providing flexibility to the generally rather thick and tough umbrella. Most coronate medusae have stiff, non-contractile tentacles that in many species are held up over the bell in healthy living specimens. Deep-water species typically possess striking dark red pigmented tissue in the stomach and other parts of the body.

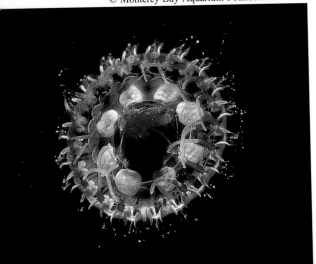

Family Atollidae

76. *Atolla vanhoeffeni*

Identification: Bell to 5 cm in diameter, usually less than 3 cm, with an obvious annular furrow some distance down from the apex. Stomach with 4 lips; base of stomach cross-shaped at the attachment point. With 20 tentacles, alternating with the same number of rhopalia, and 40 lappets. With 8 round gonads, near the base of the stomach. Usually with deep red to blackish-purple stomach, with 8 diagnostic red spots, one on each side of the 4 gastric ostea, and with 4 pigmented arcs just outside of the 8 spots forming a broken circle; gonads yellowish brown. *Natural History*: A relatively small species. The mature eggs are nearly 1 mm in diameter within the gonad. The large egg size may indicate direct development although the complete life cycle is not known. *Range and Habitat:* Probably cosmopolitan in deep water between about 500 and 1000 meters.

77. *Atolla wyvillei*

Identification: Bell to 15 cm in diameter, but usually less, with a very obvious annular furrow some distance down from the apex. Stomach with 4 lips; base of stomach shaped like a four-leaf clover at the attachment point. Usually with 22 tentacles (but may have 20-36), alternating with the same number of rhopalia and twice as many lappets. With 8 oval or bean-shaped gonads, in pairs, near the base of the stomach (male gonads may become quite elongate). Usually with a deep reddish-brown stomach and with pigmentation on the exumbrella varying from not present through several intermediate levels of reddish coloration to uniformly dark. *Natural History*: The largest and best-known of the *Atolla* species. *Atolla* swim with their tentacles trailing – one tentacle is almost always about twice as long as the others. *Range and Habitat:* Cosmopolitan; deep water, 500 to 1500 meters; in areas of upwelling it may be found nearer the surface.

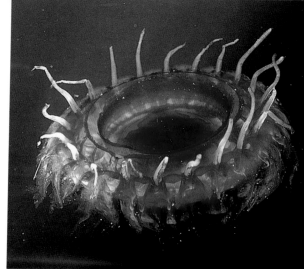

© Monterey Bay Aquarium Foundation - 1998
R. Larson

Family Linuchidae

78. *Linuche unguiculata* Thimble jellyfish

Identification: Bell to about 25 mm high; covered with nematocyst warts. Thimble-shaped, with a shallow annular furrow near the apex; stomach urn-shaped with 4 recurved lips. With 8 short tentacles and with 8 rhopalia; 16 marginal lappets. With 8 scimitar-shaped gonads. Exumbrella transparent and unpigmented, subumbrella with many dark irregular splotches on a white background, gonads yellow to brown (males) or blue-gray (females). *Natural History*: Swims nearly continuously horizontally. Feeds by capturing prey on the nematocyst-covered lappets. Prey include copepods, shrimp, chaetognaths, and fishes. The dark brown patches on the subumbrella are filled with zooxanthellae. *Range and Habitat:* Worldwide distribution in tropical and subtropical waters, usually nearshore and just below the surface in spring and summer; polyps on coral rubble.

Family Nausithoidae

79. *Nausithoe ?atlantica*

Identification: Bell to about 5 cm wide, flattened, with a thick central disk; below the annular furrow are 16 radial thickenings. With 8 tentacles alternating with 8 rhopalia; 16 marginal lappets. Mouth is simple and cruciform with 4 lips. With 8 round to triangular gonads. The exumbrella is smooth rather than pitted, and the medusa is overall a deep brick red, with stomach and gonads somewhat darker. *Range and Habitat*: Specimens collected in San Clemente Basin off San Diego and off Hawaii, in deep water below 800 m. *Remarks*: *Nausithoe punctata* is a lightly-pigmented near-surface coastal species that might be encounted in warm waters.

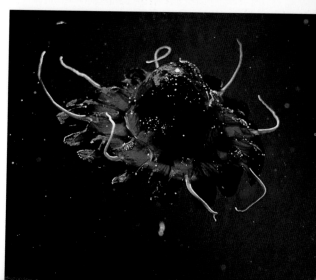

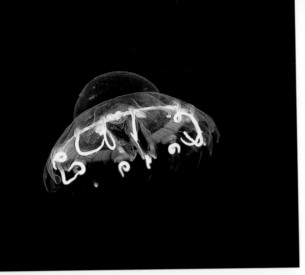

Family Paraphyllinidae

80. *Paraphyllina* spp.

Identification: Bell to 88 mm in diameter, with a very obvious annular furrow, stomach with 4 lips. With 12 tentacles, alternating in groups of 3 with 4 rhopalia; with 16 lappets. The rhopalia are in line with the 4 corners of the stomach. With 8 gonads in pairs near the base of the stomach–the gonad shapes vary from bean-shaped to elongate and curved to W-shaped. Entire exumbrella including lappets studded with nematocyst warts, and either entirely pigmented brick red or colorless with stomach and tentacles brick red. *Natural History*: A vigorous swimmer that holds its tentacles radiating out from the bell much of the time. *Range and Habitat*: In deep waters of most oceans. Three species may occur off the west coast: *P. intermedia*, *P. ransoni*, and *P. rubra*.

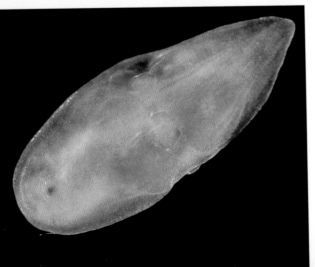

Family Periphyllidae

81. *Periphylla periphylla*

Identification: Bell to 20 cm in height with a rounded or nearly conical apex, with a very obvious annular furrow and with 16 thickenings in a ring below it. Stomach with 4 lips. With 12 tentacles, alternating in groups of 3 with 4 rhopalia; with 16 lappets. The rhopalia alternate in position with the 4 corners of the stomach. With 8 U-shaped gonads in pairs, near the base of the stomach. Stomach ususally deep reddish-brown, pigmentation variable elsewhere on the subumbrella, and with a transparent exumbrella. *Natural History*: Migrates vertically; feeds on copepods and probably other small crustaceans. Swims most of the time, holding its tentacles high. *Range and Habitat*: Perhaps the most abundant and widely distributed deep-sea scyphomedusa. In all oceans, typically below 900 meters with juveniles higher and large specimens deep, but occurring to the surface in high latitudes.

Family Tetraplatidae

82. *Tetraplatia volitans*

Identification: Roughly cylindrical, 4–9 mm long, with pointed ends and a constriction closer to one end (aboral) than the other. Oral and aboral portions of the body connected by 4 flying buttress-like structures. With 4 tracks of nematocysts running down the flying buttresses, and 4 shorter tracks of nematocysts between these. The groove contains umbrella tissue that is divided into 8 pairs of "lappets." With 8 marginal sense organs, situated in clefts between the lappets. With 4 gonads, each with paired oral and aboral lobes. Without tentacles. Color whitish or bluish white. *Natural History*: This bizarre and uncommon form might be found at any time of year. Feeds on small zooplankton. *Range and Habitat*: Worldwide, oceanic, occurring from the surface to about 900 meters.

Order Semaeostomeae

This order contains the most familiar jellyfish, including moon jellies, sea nettles and the lion's mane. All of the shallow west coast semaeostome jellies are large, conspicuous forms and easily seen even by casual observers. Jellies in this order typically have 4 or more frilly oral arms that can be quite long, and a scalloped bell margin. Most have a polyp stage that produces young medusae (ephyrae) by strobilation. With their abundance and large size, semaeostome medusae play an important role in nearshore, oceanic and deep-sea ecosystems, serving both as predators and sources of food for other organisms.

Family Pelagiidae

83. *Chrysaora fuscescens* Sea nettle
Identification: Bell to 30 cm wide, with 24 tentacles arising from the edges of the margin between lappets in groups of 3, alternating with 8 rhopalia. Oral arms long and pointed and much folded. Umbrella yellowish-brown or reddish-brown, and darkest near the margin; may or may not have a lighter 16–32 rayed star pattern on the outer portions of the exumbrella; tentacles and oral arms dark. *Natural History*: Most common in fall and winter in nearshore aggregations, but may be a near-surface slope species. Has a very unpleasant sting. Swims continuously with the tentacles and oral arms extended; feeds on planktonic crustaceans, pelagic tunicates and molluscs, ctenophores, fish eggs and larvae, and other jellyfishes. *Range and Habitat*: Common off California and Oregon and occasionally off Washington and British Columbia to the Gulf of Alaska. It has been collected as far south as Mexico.

R. Brodeur

84. *Chrysaora melanaster*
Identification: Bell to 60 cm wide, with 24–40 tentacles arising from the edges of the margin between lappets; with 8 rhopalia. Oral arms many times longer than the bell diameter, pointed and much folded. With 16 very dark radiating streaks on the subumbrella between the stomach pouches. Umbrella with a pale background; larger individuals with a reddish-brown pattern of a ring near the apex and 16 radiating streaks on the exumbrella. *Natural History*: Common in the Bering Sea where it is frequently found in association with juvenile pollock. *Range and Habitat*: Japan, Kamchatka, Aleutian Islands, and the Bering Sea. The species of *Chrysaora* have been widely confused and the southern extent of *C. melanaster* is not certain, but it does not seem to occur as far south as Oregon and California.

85. *Pelagia colorata*   Purple-striped jelly

Identification: Bell to about 70 cm diameter, covered with small nematocyst warts. With 8 marginal tentacles alternating with 8 rhopalia; 16 lappets. With 4 long oral arms that have a distinctive coiled spiral posture. With 4 interradial gonads. Small specimens pink with dark reddish tentacles, but large specimens have magenta, brown and blue colors combining on a pale exumbrella to form a deep purple pattern composed of an apical ring and 16 radiating streaks and dark colored lappets. *Natural History*: Not seen in large surface aggregations, often damaged. May be heavily infested with juvenile slender crabs (*Cancer gracilis*). Has a strong sting. Feeds on ctenophores, pelagic tunicates, fish eggs and larvae, planktonic crustaceans and other scyphomedusae. *Range and Habitat*: Bodega Bay, Monterey Bay, Santa Barbara, San Pedro Basin (California); probably oceanic or slope water.

G. Dietzmann

86. *Pelagia noctiluca*

Identification: Bell rarely exceeding 9 cm diameter and covered with nematocyst warts. With 8 hollow marginal tentacles with numerous longitudinal muscle furrows near their bases alternating with 8 rhopalia; 16 lappets. With 4 long oral arms with crenulated margins. Color variable, but often purple or yellow. *Natural History*: May occur in large numbers. The eggs transform directly into new medusae without any benthic polyp in the life cycle. Opportunistic predator that eats salps, doliolids, larvaceans, hydromedusae, ctenophores, chaetognaths, planktonic crustaceans and fish eggs. Has an uncomfortable sting. *Range and Habitat*: A common epipelagic oceanic species of warm, and sometimes temperate, seas. Sometimes seen in great numbers either in surface waters or stranded on beaches, although rarely as far north as southern California.

Family Cyaneidae

87. *Cyanea capillata*   Lion's mane

Identification: Bell up to 2 m in diameter in high-latitudes; southern specimens usually closer to 50 cm. Margin divided into 8 thick lobes, with 8 clusters of up to 150 tentacles arranged in several rows, arising from horseshoe-shaped regions between the lobes; with 8 rhopalia. Oral arms short and highly folded, forming a blocky mass only about as long as the bell is wide. Color deep brick red to purplish, yellowish-brown in small specimens. The swimming medusa looks like an 8-pointed star at the end of its power stroke. *Natural History*: Probably live less than 1 year, commonly seen in summer. Has an unpleasant sting that lasts about 4–5 hours. *Range and Habitat*: North-boreal and circumpolar; common in Alaska and Washington, occasional in Oregon, probably not in California.

Family Ulmaridae
Subfamily Aureliinae

88. *Aurelia aurita* Moon jelly

Identification: Bell to about 50 cm diameter, usually less; margin scalloped into 8 lobes, demarcated by 8 rhopalia. With 4 oral arms that meet at the center of the subumbrella (without central manubrium separating the oral arms). Oral arms extend beyond the bell usually with densely crenulated margins, but varying to smooth. With numerous fine tentacles. Usually opaque whitish, although sometimes infused with other colors. *Natural History*: Females carry clumps of developing larvae on the innermost edges of the oral arms. *Range and Habitat*: Probably not native to the west coast of North America, but has been collected in San Francisco Bay. Common in Europe, Japan, Gulf of Mexico and east coast. May be endemic to Europe and introduced elsewhere.

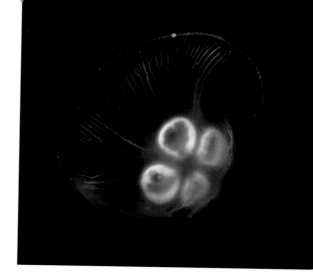

89. *Aurelia labiata* Moon jelly

Identification: Bell to about 40 cm diameter, usually less; margin scalloped into 8 major lobes that are secondarily notched so appear to be 16 lobes, with 8 rhopalia at the major notches. With a substantial, conical manubrium from which the oral arms emerge. Oral arms extend just to the edge of the bell, rarely beyond, with margins only somewhat frilled. With numerous fine tentacles. Usually opaque whitish, although sometimes infused with pink, lavender or yellow. *Natural History*: Can occur in dense aggregations mid-summer through autumn. Females carry clumps of developing larvae on the central manubrium. Males occasionally seen with long filaments of sperm trailing. *Range and Habitat*: Known from SE Alaska to Newport Beach, and Honolulu. This is apparently the "native" west coast species. *Remarks*: *A. limbata* of the Bering Sea is similar, but with a dark brown rim.

Subfamily Sthenoniinae

90. *Phacellophora camtschatica* Fried egg jellyfish

Identification: Bell to 60 cm in diameter, with 16 clusters of tentacles, slightly inside the bell margin, each containing a single row of tentacles. With 16 lappets and 16 rhopalia. Oral arms relatively short and massively folded. Central gonadal mass yellow, with surrounding clear to whitish or pale yellow bell, oral arms and tentacles, resembling a raw egg; small individuals often colorless or milky white. *Natural History*: Spends much time motionless or slowly pulsing the bell while drifting with tentacles extended 10–20 feet or more. Feeds on gelatinous zooplankton, especially other medusae. Usually with symbiotic amphipods on the subumbrella and juvenile crabs on the exumbrella. Has only a mild sting. *Range and Habitat*: Chile to Kamchatka and Japan in the Pacific; scattered distribution globally.

Subfamily Poraliinae

91. *Poralia* sp.

Identification: Bell diameter to about 40 cm; jelly soft and flaccid, exumbrella covered with small nematocyst warts. Manubrium short and tube-like, with 5–8 short oral arms. With about 20–40 broad radial canals, numerous rhopalia and tentacles. Gonads form a nearly continuous ring around the manubrium. All surfaces brownish red, except for the white rhopalia. *Natural History*: Often occurs near the bottom, sometimes dragging its tentacles. Submersible observations have revealed that this medusa is not uncommon, but the near-bottom habitat means that it is infrequently collected. Swimming contractions are so slow that it seems to be moving in slow motion. *Range and Habitat*: Probably worldwide mesopelagic; two undescribed species have been distinguished.

P. Dayton

Subfamily Stygiomedusinae

92. *Stygiomedusa gigantea*

Identification: Up to at least one meter in bell diameter, jelly thick and soft, shaped like a broad-brimmed ladies hat, thicker in the center and thinning towards the margins. With four long slender oral arms that extend many meters. With 20 straight radial canals alternating with 20 branched radial canals near the margin; the branching of the radial canals forms a general reticulum. Without tentacles; with 20 rhopalia. Colored a deep brick red all over, including throughout the mesoglea; gastrovascular epithelium whitish. *Natural History*: Said to be viviparous, one specimen contained two fully formed, 10 cm diameter juveniles. *Range and Habitat*: Mesopelagic, probably cosmopolitan, and occasionally near the surface at high latitudes (photo taken in Antarctica); in deep trawls off Santa Barbara and San Diego, and Endeavor Ridge, Washington.

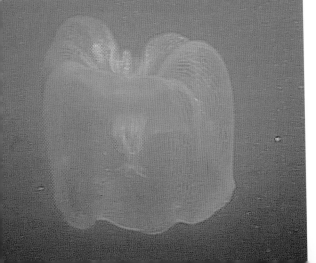

Subfamily Deepstariinae

93. *Deepstaria enigmatica*

Identification: To at least 50 cm in diameter, but may be much larger; very flimsy and malleable with uniformly thin, soft jelly. Stomach is a short tube with 5 narrow oral arms. Radial canal system is an elongate meshwork over the entire umbrella, with increasingly smaller mesh size near the margin. Without tentacles; with numerous (20?) rhopalia. Colorless or yellowish brown, with brown canals and/or stomach. *Natural History*: Probably fewer than 10 specimens of this genus have been collected. Direct observation from submersibles shows that it is rather cylindrical when relaxed, but can rapidly close the bell margin, and then contracts as a peristaltic wave moves from top to bottom. *Range and Habitat*: Probably cosmopolitan; has been collected off Oregon, California, and Hawaii between 600 and 1750 meters.

Order Rhizostomeae

Rhizostome medusae typically inhabit shallow tropical and subtropical seas. A few species may be seen in the southern part of the range of this book. Rather than the oral arms and central mouth of semaeostome medusae, they possess a system of 8 branched, lobed, arm-like appendages. The edges of the fused lobes bear many tiny mouth-like openings. Lacking marginal tentacles, rhizostome jellies feed on tiny organisms using the mouth lobes. Many tropical species also harbor zooxanthellae in the tissue, which may supplement their nutrition. Most jellyfish harvested for human consumption have been rhizostome species.

R. Larson

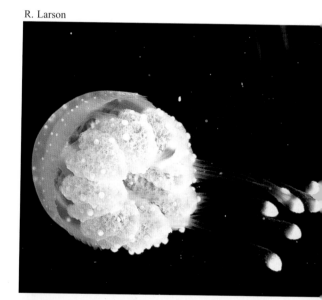

Family Mastigiidae

94. *Phyllorhiza punctata*

Identification: Bell to 50 cm wide with stiff jelly; umbrella finely granular. Without marginal tentacles, but with 8 long, fleshy oral arms that hang below the bell. Each oral arm is three-winged, with large openings in the lateral membranes and bluntly-ending filaments on the lower portion. With 8 marginal sense organs and up to 14 lappets per octant. Umbrella bluish or brownish with white spots, oral arms with blue and white tips. *Natural History*: Active, strong swimmers, they capture small zooplankton on small digitate filaments on the lower surface of the oral arms. Planulae are attached to the arm disk filaments of the females and can be settled in the laboratory; polyp not known from the field. *Range and Habitat*: An Indo-Pacific species that has also been found in Hawaii and the western tropical Atlantic (where it appears to have been introduced in the 1960s or early 1970s); occasionally seen in southern California.

R. Larson

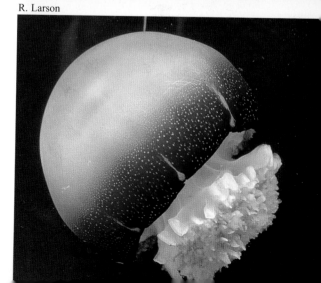

Family Stomolophidae

95. *Stomolophus meleagris*　　Cannonball jellyfish

Identification: Bell to 18 cm wide, jelly thick and rigid; without marginal tentacles. With 8 oral arms mostly coalesced into roughly a cylinder, but branching complexly at the distal ends, forming a stiff feeding structure that hangs slightly below the bell margin. With 8 marginal sense organs and about 14 lappets per octant. Bell bluish or yellowish, becoming darker towards the margin and speckled with blue spots; oral arms pale blue, lips whitish. *Natural History*: Strong active swimmers, frequently occur in swarms. Collect small zooplankton prey. Can be eaten after treatment with salt and alum. Polyps not known from the field. *Range and Habitat*: Coastal, during the summer; warm Atlantic, Caribbean, and Gulf of Mexico, but occasional on the Pacific coast from San Diego and Gulf of California to Equador.

Phylum Ctenophora

Ctenophores, also known as comb jellies, are a remarkable and distinctive group of gelatinous predators. At times planktonic comb jellies can be very abundant and ecologically important. Some occur in coastal habitats or near the ocean surface, while many species are found only in the deep open ocean. Comb jellies are unique in having eight rows of ciliated plates (the comb rows) which are used for locomotion. With shimmering rainbow patterns formed by the comb rows, ctenophores surely rank among the living jewels of the sea. Although ctenophores resemble cnidarian jellies in that they are carnivores, are often transparent, and have a gelatinous consistency, they are not closely related and lack the alternation of polyp and medusa generations found in most pelagic cnidarians.

Ctenophores are the largest animals to use cilia for locomotion. The cilia are arranged into comb plates, or ctenes, that are lined up to form the comb rows. Comb plate cilia may be up to 2 mm long. The power stroke of each plate is produced by a rapid swing of the straightened cilia, followed by a slower, curled return stroke. A metachronal wave starts at the aboral end to push the animal with its mouth in a forward direction. Each comb plate is triggered to begin the beat by movement of the adjacent plate. When light strikes the rows at the proper angle, a beautiful shimmering iridescence results from the diffractive effects of the closely-spaced cilia. Most ctenophores can also produce intrinsic bioluminescent light through the action of calcium-activated photoproteins. The faint light produced by these compounds can only be seen in a darkened area and is not visible under normal viewing conditions. The colorful patterns formed when light strikes the comb rows (often seen in photographs) have nothing to do with bioluminescence.

A statocyst aids in the orientation of comb jellies. This consists of a statolith (a tiny, calcareous grain) in contact with ciliated receptor cells. Each row of combs is connected to the statocyst by a minute, ciliated groove that coordinates beating between rows. Position changes by the ctenophore result in mechanical stimulation of cilia in the statocyst, causing asymmetrical beating of comb row cilia that keep the animal in its proper orientation. When changes in salinity are encountered, at least some ctenophores can adjust their buoyancy, probably through passive osmotic accommodation. Since most ctenophores are oceanic, however, they generally do not tolerate large changes in salinity. The active exclusion of sulfate ions at the cellular level assists in establishing species-specific equilibrium buoyancy.

With the limited escape capability and minimal chemical defenses of ctenophores, transparency is an important means for avoiding predation. Many cnidarian jellies prey on ctenophores. Ctenophores are predators on various types of zooplankton, including some types that prey on other comb jellies. Food is brought to the mouth by a variety of mechanisms, digested in the pharynx (stomach), and transported throughout the body by complex systems of canals. Ingestion rates for lobate and cydippid comb jellies generally increase linearly with increasing food concentration, although digestive efficiency tends to decline. At high food densities these groups may thus have an advantage over other zooplankton predators, such as chaetognaths. Ctenophores often serve as hosts for various types of hyperiid amphipods, which eat the host's food or tissue depending on availability. Certain visual predators, such as fish and sea turtles, eat comb jellies.

Most ctenophores are simultaneous hermaphrodites, and those tested appear to be capable of self-fertilization. However, at least one family, the Ocyropsidae, is dioecious, having separate sexes. The gonads are located beneath each comb row on the walls of the meridional canals. Eggs and sperm are released directly into the water, with external fertilization of eggs that develop into ciliated larvae.

Order Cydippida

Cydippid comb jellies are characterized by a solid spherical or ovoid body, with a pair of long, retractable tentacles arising within sheaths on opposite sides of the body. The tentacles generally can be withdrawn completely into the sheaths, and are present in both the larval and adult stages. Numerous sticky colloblast cells on the tentacles and side branches (also known as tentilla) are used to capture prey. When prey are caught and transferred to the mouth, fishing activity must be interrupted. A long pharynx, or stomach, connects to a complex gastro-vascular and meridional canal system, with branches that end blindly. The aboral end houses the complicated apical sense organ.

Family Haeckeliidae

96. *Haeckelia beehleri*

Identification: Usually under 10 mm in length, elongate body nearly circular in cross section. With long tentacular sheaths, and unbranched tentacles exiting near the mouth. Mouth wide and pharynx large. The 8 comb rows are nearly equal in length and short, extending from the aboral pole to less than the middle of the body. With 8 canals underlying the comb rows, equal in length and extending nearly the entire body length. Body translucent or slightly opaque, sometimes light tan colored. *Natural History*: Has a very malleable body, making frequent minor shape changes. *Range and Habitat*: Surface off Santa Barbara; Atlantic Ocean and the Mediterranean. Probably widespread, but uncommon.

R. Harbison
S. Haddock

97. *Haeckelia bimaculata*

Identification: Very small, usually less than 3 mm with ovoid appearance and ellipsoid cross section. Tentacle sheaths long, with unbranched tentacles exiting about 1/4 body length below mouth. Mouth distinctly wide and pharynx large, with only 4 digestive canals, each extending 3/4 the body length. The 8 comb rows are nearly equal in length and extend from the aboral pole to the middle of the body. Body fairly transparent, with large orange-red pigment spots on the canals around the base of the pharynx, infundibulum and tentacle bases, and small red spots along the comb rows. *Natural History*: Eats narcomedusae and like *H. rubra*, carries narcomedusan nematocysts in its tentacles instead of colloblasts. It requires a careful eye to find this little ctenophore. *Range and Habitat*: Surface waters off Santa Barbara and in the Mediterranean. Probably more widely distributed.

a. S. Haddock   b. C. Mills

98. *Haeckelia rubra*

Identification: Small, usually under 7 mm length, elongate with a narrowed oral end. Tentacle sheaths long; unbranched tentacles exit near the mouth. Mouth distinctly wide and pharynx large, with only 4 digestive canals, each extending 3/4 body length. The 8 comb rows are unequal in length and extend from the aboral pole about 3/5 of the body length. Body transparent and frequently tinted green, with two pairs of orange-red pigment spots, on the tentacle bases and the tentacle sheaths. *Natural History*: Preys on narcomedusae, such as *Aegina*, *Solmaris* and *Solmundella*, storing their nematocysts in its tentacles. Can also eat planktonic crustaceans. *Range and Habitat*: Infrequently collected in surface waters at Friday Harbor, in British Columbia fjords at depths below 100 meters, and near the surface off Santa Barbara; also Mediterranean, Japan, and China. Probably cosmopolitan.

Family Bathyctenidae

99. *Bathyctena chuni*

Identification: Up to 40-50 mm long, slightly flattened. Oral end dominated by broad, nearly semi-circular mouth; aboral end bluntly rounded. Comb rows narrow and evenly spaced, reaching about 1/2 the distance from aboral end to the mouth. With 8 rows of yellowish-white spots alternating with the comb rows. Tentacle bases lie near the pharynx, with tentacle shafts angling upward toward the corners of the mouth, where they exit as 2 small holes. Tentacles with many fine side branches. Body milky white and fairly opaque; inside of the mouth is darkly pigmented – carmine red to deep purple to nearly black. *Natural History*: Lives in very deep water. Trawl-collected specimens may turn inside out as dark red balls. The rows of spots exude bioluminescent material. *Range and Habitat*: Probably cosmopolitan in deep water from 1000–3500 meters.

Family Lampeidae

100. *Lampea pancerina*

Identification: Body to several cm long, circular in cross section, tapered toward the oral end. Comb rows of equal lengths, extend from the aboral end to about the opening of the tentacle sheaths. Very large and voluminous pharynx, occupying 4/5–5/6 of the body length, with a highly extensible mouth. Tentacle bulbs short and near the midline of the body, with sheaths exiting horizontally near the midline or toward the aboral pole. Tentacles with relatively few short side branches. Milky white in color, or with a rosy tinge and pink pigmentation in the comb rows. *Natural History*: Apparently feeds exclusively on salps. Small individuals attach to salps, flatten out and resemble parasites; large individuals can engulf entire salp chains. *Range and Habitat*: Uncommon in Monterey Bay and deep water off Santa Barbara; best known from Mediterranean, Atlantic Ocean and Japan.

Family Pleurobrachiidae

101. *Hormiphora californensis*

Identification: Body about 30 mm in length, oblong, moderately compressed. Broadly rounded to flattened at the aboral end; small specimens narrower at the oral end. Comb rows equal, extending from near the aboral pole to at least 4/5 the distance to the mouth. Canals underlying the comb rows without branches. Mouth relatively small; pharynx extends about 2/3 body length. Tentacle bulbs are about 1/3 the body length, located midway along the body and close to the pharynx; the very long sheaths parallel the pharynx, exiting the body near the aboral end. Tentacle side branches numerous. Body transparent, with yellow brown tentacles. *Natural History*: Like *Pleurobrachia bachei*, this cydippid is not luminescent. Capable of eating euphausiids. *Range and Habitat*: At least Santa Barbara to San Diego, to 200 meters.

102. *Hormiphora* undescribed species

Identification: Body usually less than 35 mm long, ovate. Circular in cross section, somewhat flattened at the aboral end and narrowed towards the mouth. Comb rows equal, extending about 3/4 distance from the aboral end toward the mouth. The canals underlying the comb rows extend nearly to the mouth and have branches pointing inward toward the pharynx. Pharynx extends for more than half the body length. Tentacle bulbs large, 1/3–1/2 the body length, close to the pharynx, nearer to the mouth than the aboral pole; the large tentacle sheaths lying close to the pharynx, exiting the body near the aboral end. Tentacle side branches numerous. Body transparent and colorless. *Natural History*: Similar to *Pleurobrachia*, but usually deeper in the water column and typically with a larger, distinctly more elongate body. Appears to move up from below 100 m to shallower water at night. *Range and Habitat*: Known from deep water in Monterey Bay; occasionally seen in Friday Harbor.

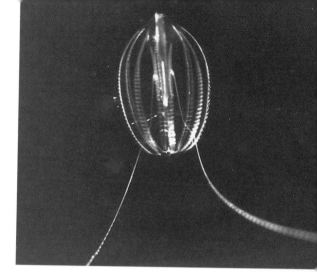

103. *Hormiphora cucumis*

Identification: To at least 10 cm in length and cucumber-shaped, slightly flattened. Comb rows and their underlying canals run nearly the entire length of the body and are all equal in length. Tentacle bulbs are long and narrow, running very near the pharynx, measuring more than 1/4 total body length. Tentacle sheaths run a short distance toward the aboral end and then angle abruptly outward, about 1/3 body length from the aboral pole. Mouth large, pharynx occupies just over 1/2 of the body's length. Tentacles long, with side branch filaments all alike, spaced a few mm apart on a fully extended tentacle. Very transparent and unpigmented or slightly bluish. *Range and Habitat*: Originally described from the central Gulf of Alaska and ocean off Tokyo; also collected infrequently in Prince William Sound, Friday Harbor, and upper 10 m of the open ocean off the border between Washington and British Columbia. Probably oceanic, rather than coastal; strong swimmer.

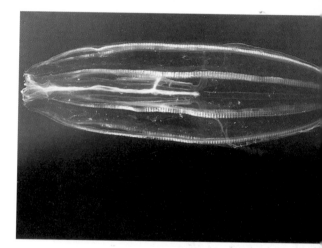

104. *Hormiphora palmata*

Identification: To several cm in length, soft and flaccid, ovate or spindle-shaped, larger specimens with a long, tapering oral end, slightly flattened; comb rows run 2/3–4/5 of the distance up from the aboral pole, are evenly spaced and all the same length. The canals underlying the comb rows extend beyond the ctenes, but end several mm short of the mouth. The tentacle bulbs are very long and narrow, running horizontal to, and very near the pharynx, measuring about 1/3 or more of the total body length. The voluminous tentacle sheaths angle towards the aboral pole and exit about 1/3–1/5 body length or less from that pole. Mouth large, pharynx occupies more than 1/2 of the body's length. Tentacles long, with side branch filaments simple and all alike. Transparent and colorless, or with yellow tentacles. *Range and Habitat*: Known from surface collections, Hawaii and Mexico; oceanic, warm water.

105. *Pleurobrachia bachei*   Sea Gooseberry

Identification: Body nearly spherical, or with a narrowed mouth, to about 15 mm. Comb rows evenly spaced, extend nearly the entire length. Mouth small, pharynx usually less than half the total body length. Tentacle bulbs short and pressed closely against the pharynx, tentacle sheaths angle down and open in the lower 1/4. Tentacles with numerous side branches. Body transparent, with yellowish red on the tentacles and sometimes with purplish blotches near the opening of the pharynx. *Natural History*: Acts as an ambush predator, setting its tentacles by swimming upward in a spiral; the body spins to bring a food-laden tentacle to the mouth. Preys on copepods, larval fish, various types of eggs and other small plankton. Often harbors hitchhiking hyperiid amphipods. Unlike most ctenophores, *Pleurobrachia* lacks any kind of bioluminescence. *Range and Habitat*: SE Alaska to Acapulco, Mexico; very common.

Family Euplokamidae

106. *Euplokamis dunlapae*

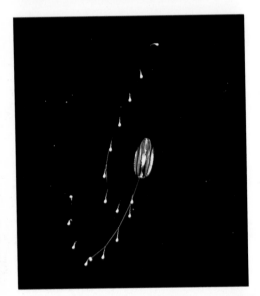

Identification: Length to 20 mm. May have gelatinous extensions protruding below the statocyst. Nearly circular to slightly flattened in cross section. Comb rows lie along 2/3–3/4 of the body length and are all the same length. Tentacle bases parallel to the pharynx, mid-body, about 1/4 the body length. Tentacles exit about 1/4 the body distance away from the aboral pole with 10–60 side branches that are normally held tightly coiled. Transparent, with red pigment at edges of the comb rows, tentacle bases and tentacle side branches. *Natural History*: Feeds on both small and large copepods by wrapping the prey with its tentilla. Exudes bioluminescent globules in its wake when disturbed. *Range and Habitat*: San Juan Islands and fjords of British Columbia, usually 100 to 600 meters deep, but occasionally at the surface; Gulf of Maine. A little-known, but abundant midwater ctenophore.

S. Haddock

Family Mertensiidae

107. *Charistephane fugiens*

Identification: Body to about 20 mm, oval, highly flattened. Comb rows short, ending below the midline, with only a few large comb plates (no more that 7 ctenes per comb row). The canals underlying the comb rows extend to the oral end. Pharynx small, extending only 1/4 of the body length. Tentacle bulbs small and angled obliquely, at the midline of the body and very near the outside walls. Tentacle sheaths short, exiting the body towards the aboral end as funnels beginning just below the midline. Tentacles with short, widely-spaced side branches. Extremely transparent. *Range and Habitat*: Deep midwater off Santa Barbara, San Clemente Basin, and Hawaii. Originally described from the Mediterranean. Poorly known.

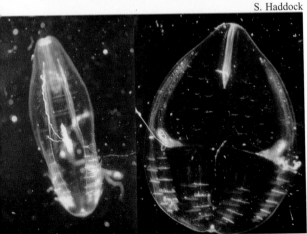

108. Undescribed species of mertensiid

Identification: Body to 20 mm, ovate, laterally compressed. Comb rows equal in length, extending about 4/5–5/6 of the entire body, and raised up on ridges. Tentacle bulbs (1/3–1/2 body length) near center of the body. Tentacles have enormous numbers of very fine side branches, exit through long slit-like openings that run down most of the aboral half of the animal. Body colorless to translucent. Tentacle bulbs and tentacles with claret or rose-violet pigmentation. *Natural History*: Feeding animals static with mouth up, extending their tentacles to form a diaphanous net that looks like a mucous sheet. Eats amphipods and copepods, although the wide-opening mouth suggests it may take larger prey. *Range and Habitat*: From 100-600 meters in British Columbia fjords, occasionally near the surface in Friday Harbor in the winter, near the surface in Monterey Bay and Santa Barbara.

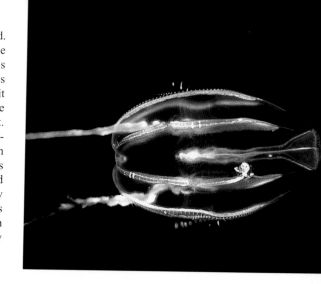

Family Dryodoridae

109. *Dryodora glandiformis*

Identification: Body acorn-shaped, nearly cylindrical in cross-section, with length to 5 cm in polar regions, but rarely exceeding 10 mm at Friday Harbor. Tiny tentacle sheaths lie near the midline of the body and are not large enough for complete retraction of the fine, unbranched tentacles. Comb rows extend about 1/2 the body length. Oral end is not actually the mouth but is the opening to a vestibular area in which prey is held prior to being captured by the mouth deep inside. Transparent, with red pigment in the tentacle sheaths, sometimes with reddish-brown pigment along the meridional canals and oral pole. *Natural History*: Oral end opens very wide; feeds on appendicularians. Eggs emerge in strings of 20 to 30. *Range and Habitat*: Boreal waters of both hemispheres; known from Friday Harbor and British Columbia, where specimens are much smaller and more transparent than polar individuals.

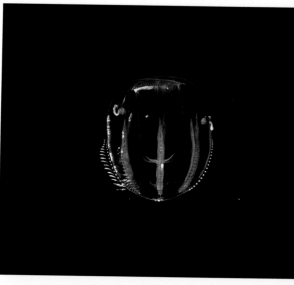

Order Thalassocalycida
Family Thalassocalycidae

110. *Thalassocalyce inconstans*

Identification: Body shaped like a broad medusa, up to 15 cm when fully expanded. Slit-like mouth on a central conical peduncle. Two small tentacle bulbs situated on the sides of the peduncle, and the tentacles (with short side branches) hang into the "bell" cavity. Comb rows are short and on the upper surface. Transparent and colorless, highlighted by looping patterns of the digestive canals running through the jelly. *Natural History*: May resemble a hydromedusa. Feeding behavior is similar to lobate ctenophores. Bell is fully expanded when feeding and contracts when the animal is disturbed or captures prey. Has no escape response and only limited swimming ability. *Range and Habitat*: Atlantic Ocean, Bahamas, Mediterranean, and several locations off California, surface to midwater.

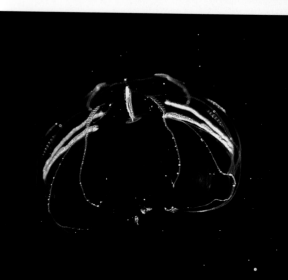

Order Lobata

Lobate comb jellies have a pair of distinctive large lobes at the oral end of the body. Most have exceedingly delicate tissue that is easily damaged. Tentacles are generally small and may appear to be either a pair of branched tentacles in sheaths near the mouth, or as numerous small filaments in tracts leading toward the mouth. The tentacles are held in a cavity enclosed by the oral lobes. Lobate ctenophores possess four ciliated, ribbon-like auricles, two on each side of the body above the mouth. The mouth is wide and continues into an oral groove extending to the base of the oral lobes. The mucus-covered oral lobes are used to capture small zooplankton prey. Feeding is a continuous process and is not interrupted by transfer of prey to the mouth (as in cydippids). Larvae pass through a tiny cydippid-like stage with tentacles. Lobes and auricles develop during metamorphosis into the adult form as the tentacles become reduced.

Family Bathocyroidae

111. *Bathocyroe ?fosteri*

Identification: With broad oral lobes that are used for locomotion; measuring up to 4 cm across the lobes. Short comb rows at the aboral end near the statocyst. With broad, flat auricles. Tentacle bulbs situated on either side of the gut, with slender tentacles lying along the mouth opening. Body transparent or translucent white with readily visible patterns of digestive canals on the lobes, and distinctive red gut. *Natural History*: Submersible observations reveal that individuals usually hang motionless, but can swim by flapping the lobes together in a sort of frogkick. Prey include copepods, amphipods and euphausiids. Low oxygen consumption compared to many ctenophores. *Range and Habitat*: Deep cold midwater, from about 200–1000 meters, common almost everywhere that researchers have used submersibles midwater. It is not yet clear if more than one species exist.

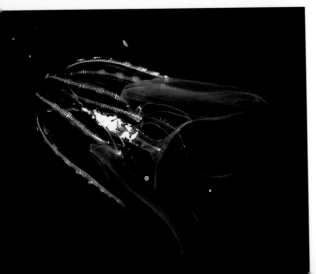

Family Bolinopsidae

112. *Bolinopsis infundibulum*

Identification: Oblong rounded body to 15 cm, with medium-sized lobes. With short, flat auricles. With 8 comb rows, unequal in length, but all 1/2–2/3 of the body length. Both sides of the mouth have a row of small, thin tentacular filaments, with a pair of longer medial tentacles at the corners. Transparent and colorless except for rows of darkly pigmented spots on the lobes that may coalesce into dark, pigmented lines. *Natural History*: Typically oriented vertically, slowly cruising up or downward, using ciliated mucous-covered oral lobes to capture copepods, euphausiids and other small zooplankton. Eaten by *Beroe*. *Range and Habitat*: Temperate and Arctic waters of Atlantic and Pacific to southern California. Small specimens usually found near the surface, with larger individuals to 1000 meters.

Family Deiopeidae

113. *Deiopea kaloktenota*

Identification: Body rounded to oval, to 45 mm in length and highly flattened. With relatively small oral lobes and 4 stout auricles. With 8 unequal comb rows, each possessing up to 30 widely-spaced comb plates. With numerous fine tentacular filaments along the edge of the mouth. Highly transparent and colorless. *Natural History*: Weak swimmer; it readily eats copepods, but has also taken *Pleurobrachia* and calycophoran siphonophores in the laboratory. *Range and Habitat*: Oceanic. Uncommon in Monterey Bay below 300 meters and at the surface off Santa Barbara; Mediterranean. *Remarks:* May prove to be a juvenile form of *Kiyohimea*, in which case the genus name *Deiopea* has priority.

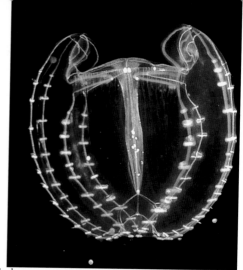

S. Haddock
© Monterey Bay Aquarium Research Institute - 1998

114. *Kiyohimea* spp.

Identification: To 30 cm or more in length, highly flattened; very fragile. Body semicircular to V-shaped, with large oral lobes having highly meandering canals and with 2 triangular or elongate aboral extensions; with 4 auricles at the bases of the oral lobes. With 8 comb rows, 4 of which extend down the 2 aboral processes; comb plates widely spaced. With a tentacle bulb on either side of the mouth, with or without oral tentacles. Transparent and colorless. *Natural History*: Submersible observations show that *Kiyohimea* tends to drift with the water mass and exhibits very little active movement. The ctenes beat only occasionally. Probably a passive ambush predator, capturing prey including euphausiids on the mucus of its lobes. *Range and Habitat*: In deep water at least off central and southern California and elsewhere in the North Pacific Ocean in the east China Sea and off Hawaii. Also in the subtropical North Atlantic Ocean. Perhaps cosmopolitan.

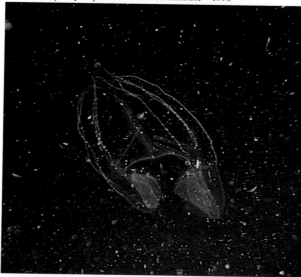

Family Leucotheidae

115. *Leucothea pulchra*

Identification: Oblong body to 25 cm or more, covered with distinctive brownish-orange papillae; extremely delicate. With oral lobes up to full body length when extended with complex patterns of meandering canals, and 4 long, worm-like auricles. With 8 comb rows, unequal in length. Oral canal lined by numerous fine tentacular filaments, and with a pair of long secondary tentacles usually trailing aborally. *Natural History*: The largest and most delicate of all the shallow lobates. When feeding, it swims slowly in a horizontal plane. Feeds primarily on copepods and other small zooplankton. Upon contact with prey, the lobe folds into a tube and brings the captive into contact with the oral tentacles for transfer to the mouth. *Range and Habitat*: Central California to Sea of Cortez, 0–200 meters.

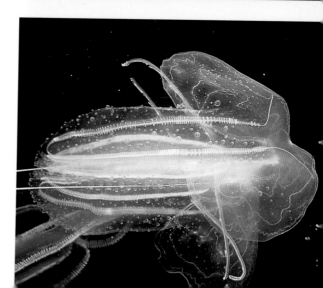

Family Ocyropsidae

116. *Ocyropsis maculata*

Identification: Body to several cm, laterally compressed and reduced, with very large, muscular oral lobes. With 8 short comb rows, extending from the statocyst to the bases of the lobes, and having relatively few comb plates; with 4 ribbon-like auricles. Pharynx hourglass-shaped. With gonads only on the portions of the canals that extend into the lobes. Transparent and colorless except for a pair of large, diffuse, dark spots usually present on the inner surfaces of the each oral lobe. *Natural History*: Forages by swimming horizontally with the ciliated auricles held rigid. Captures prey on the lobes, including cydippid ctenophores, siphonophores, pteropods, euphausiids, large copepods and fish larvae. Can swim backwards by clapping the lobes together; a cloud of luminescent mucus is released at the same time. *Range and Habitat*: Cosmopolitan in tropical and subtropical epipelagic waters; to southern California in some El Niño years.

G. Dietzmann

Order Cestida

Cestid comb jellies have highly transparent bodies flattened in the tentacular plane and greatly elongated in the pharyngeal plane, resulting in the overall shape of a belt or ribbon. The mouth is positioned at the center of the body on the leading edge of the ribbon-form, with tentacles streaming back over the body surface. Comb rows run along most of the aboral edge of the body. Brown or yellow pigment spots are sometimes present on the tips of the body. Like their close relatives the lobate ctenophores, cestids pass through a larval cydippid stage with tentacles.

G. Dietzmann

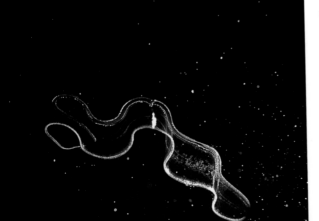

Family Cestidae

117. *Cestum veneris*   Venus' Girdle

Identification: Shaped like a wing or ribbon, up to nearly 1.5 m in length, but usually less than 80 cm. A set of canals derive from the base of the pharyx, rapidly turn up and then run lengthwise out the midline of the ribbon. Gonads in a continuous (not broken) line. Tentacles are held in a groove along the oral edge and possess numerous tentilla that trail back over the body. Transparent; may acquire a violet color or yellow pigment on the tentacles, canals, and near the tips. *Natural History*: Forages by swimming horizontally, with the oral edge leading, like a flying wing. Captures copepods and other small crustacean prey. For escape, the body undulates rapidly in a direction perpendicular to normal swimming. Tissue very delicate. Bioluminesces along meridional canals when disturbed. *Range and Habitat*: Seen occasionally in surface waters of Monterey Bay, Santa Barbara, San Diego, and the Sea of Cortez. Typically found in tropical and sub-tropical waters of all oceans.

118. *Velamen parallelum*

Identification: Ribbon-like body, less than 20 cm long. Distinguished from *Cestum* by the canals that run lengthwise out the midline of the ribbon, proceeding straight out from their points of origin in *Velamen*, whereas they turn rapidly upward to arrive at the center of the body of *Cestum*. The gonads have a characteristic broken, or dashed, appearance. Transparent and colorless or faintly yellow colored. *Natural History*: Swimming behavior like *Cestum*, but more active with frequent changes of direction; with vigorous wriggling escape mode that has been described as "darting." Densities of several hundred per cubic meter have been reported in the Sea of Cortez. *Range and Habitat*: Cosmopolitan in tropical and subtropical oceanic waters to at least 200 meters. Occasionally seen in surface waters as far north as central California.

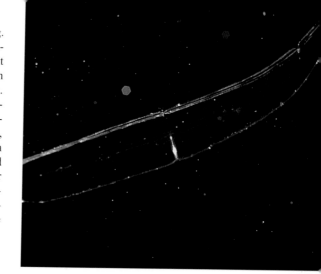

Order Beroida

Unlike the tentaculate comb jellies, members of this order have a body resembling a sac, without tentacles or lobes at the oral end. *Beroe* swim nearly all of the time in search of prey, leading with the mouth, which is normally closed in order to maintain a streamlined profile. The mouth opens to swallow prey directly, or can "bite" pieces of prey too large to swallow by using the macrocilia that line the upper region of the pharynx. These macrocilia are each formed by large numbers of cilia bound together by an outer membrane. Eight meridional digestive canals, which usually have branching diverticula, lie just below the comb rows. *Beroe* larvae lack tentacles and do not go through a cydippid stage of development; young *Beroe* of all species are typically spotted, older animals become either colorless or more pigmented.

Family Beroidae

119. *Beroe abyssicola*

Identification: Oblong with both ends rounded; length to about 7 cm; slightly flattened in cross section. With 8 comb rows of about equal length, extending 1/2–2/3 of the distance from the aboral pole toward the mouth. Meridional canals with many branches; the branches do not anastomose. Pharyngeal canals distinctly branched. Somewhat opaque, larger specimens have a pharynx lined with an intense red, purple or nearly black color, smaller specimens may be nearly colorless. *Natural History*: The deep rose coloration of the body easily distinguishes this species, but when individuals from deep water are brought to the surface, they usually lose their coloration within a few days. This color has been shown to be from symbiotic flagellates embedded in the pharyngeal lining, and hence this species may turn out to be a variant of *Beroe cucumis*. *Range and Habitat*: Usually in deep water from British Columbia to Santa Barbara.

120. *Beroe ?cucumis*

Identification: Oblong with both ends rounded; length to 6–10 cm, but typically smaller; slightly flattened to nearly circular in cross section. With 8 comb rows of two lengths, extending about 1/2 and 2/3 of the distance from the aboral pole toward the mouth. Meridional canals with some branches, the branches do not anastomose. Fairly transparent with golden tan and/or orangish-pink pigment or colorless. *Natural History*: A relatively strong, rapid swimmer that typically forages by swimming in a spiral pattern. Feeds primarily on lobate and cydippid ctenophores, also on other *Beroe* and salps. Freshly captured prey can be easily seen and identified through the body wall. Large aggregations can be found at the surface in Central California. *Range and Habitat*: Common offshore along most of the west coast near the surface and to about 500 meters. (Not confirmed to be the same as the boreal *B. cucumis*.)

121. *Beroe forskalii*

Identification: Shaped like a compressed cone up to 15 cm long, aboral end pointed and oral end very broad; highly flattened in cross section. With 8 comb rows of nearly equal length running about 3/4 of the distance from the aboral pole toward the mouth, with high numbers of very densely packed comb plates. Meridional canals with dense networks of many branches that anastomose with one another. Fairly transparent, often tinged with rose and with darker red along the comb rows. *Natural History*: A strong swimmer that forages in a spiral pattern and engulfs its prey whole (lobate and cydippid ctenophores). The mouth opens very wide and this *Beroe* can expand to huge spherical dimensions when full. Bioluminescent in the anastomosing canals. *Range and Habitat*: Widely distributed in the Atlantic, Mediterranean and Pacific; occurs in Monterey Bay and off Santa Barbara near the surface; rare in Friday Harbor. Probably oceanic.

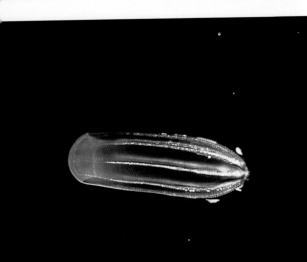

122. *Beroe gracilis*

Identification: Slender and elongate, with both ends evenly rounded; length to about 30 mm, frequently much less; slightly flattened in cross section. With 8 comb rows of equal length, about 2/3–3/4 of the distance from the aboral pole toward the mouth. Meridional canals with only a few branches that mostly extend inward; the branches do not anastomose. Mostly transparent, with reddish-brown pigment along the comb rows, near the aboral statocyst, and sometimes lining the pharynx. *Natural History*: Feeds voraciously on *Pleurobrachia*, *Bolinopsis* and siphonophores; can bite off chunks if the entire item cannot be consumed. May occur in dense swarms at the surface. A fast swimmer. *Range and Habitat*: Monterey Bay, surface off Santa Barbara and at Friday Harbor; deep water in the Bahamas; probably worldwide in cold water.

123. *Beroe mitrata*

Identification: Oblong, soft and flaccid, length to 6 cm or more; highly flattened in cross section. With aboral end round to slightly tapered and oral end very broadly rounded with nearly semi-circular mouth. With 8 comb rows of two lengths (to about the midline of the body and 2/3–3/4 of the distance from the aboral pole toward the mouth). Meridional canals with many branches that all bend towards the mouth; the branches do not anastomose. Mostly whitish or pink, somewhat opaque, with a diffuse orange or red spot or pair of stripes on either side of the body overlying the pharynx. *Natural History*: Extemely flexible and can even turn itself inside-out. It readily eats *Bolinopsis*. *Range and Habitat*: Has been seen at Friday Harbor, Japan, south central Pacific, the Mediterranean, and off South Africa. Appears to occur near the surface worldwide.

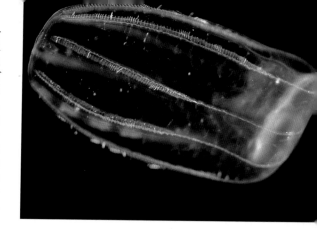

Phylum Mollusca / Class Gastropoda

Gastropods include the familiar snails and slugs that plod along on land and in the sea. A few groups, however, have taken up a planktonic existence, either at or below the surface. Some have undergone little change from their benthic relatives and lack a truly gelatinous tissue. Most have become superbly adapted to pelagic conditions with reductions or loss of the heavy shell, possession of swimming structures derived from the foot, and development of buoyant gelatinous tissue.

Subclass Prosobranchia
Order Mesogastropoda

Prosobranch gastropods include the vast majority of marine snails, primarily in the Order Mesogastropoda. They undergo the unique gastropod characteristic of torsion, which brings the gills, anus and mantle cavity toward the front of the body. Mesogastropods have a single gill (ctenidium), which is attached to the wall of the mantle cavity. The sexes are separate with males usually possessing a penis. While most are benthic grazers or predators, two groups, the janthinids and heteropods, have taken up a planktonic predatory existence.

Suborder Ptenoglossa
Family Janthinidae

R. Gilmer

124. *Janthina janthina*

Identification: Thin, smooth shell, length to 4 cm, has a typical gastropod coil. Shell and body with violet to blue color. Head with an eversible snout and pair of forked tentacles. Radula lacks a median tooth, but with a number of long, hooked lateral teeth. Foot large and conspicuous, and lacks an operculum in the adult. *Natural History*: Incapable of swimming, individuals hang suspended from the air-water interface by constructing a raft of air bubbles. Prey (typically *Velella* but also siphonophores) are grasped with the teeth. Prey nematocysts appear to have little effect on its feeding activities. *Range and Habitat*: Tropical and semi-tropical oceanic waters; relatively common as far north as southern California, and may occasionally be carried to British Columbia.

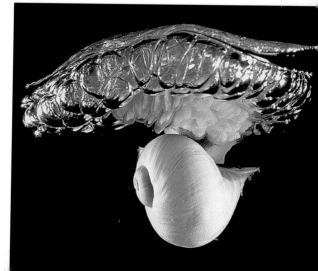

Superfamily Heteropoda

The heteropods are distinctive visual predators. Along with janthinids, they are the only prosobranch gastropods to live entirely in the water column. About 30 species form this group of highly modified, mostly warm water, open ocean dwellers. Planktonic life is made possible by a great reduction in the shell, presence of a medial swimming fin, and a largely transparent body. Undulation of the fin (which is held upward) propels the animal. A pair of well developed eyes aids prey capture. The radula, with prominent chitinous teeth, is held at the end of a long proboscis. A conspicuous penis is present in males, and fertilization is internal via spermatophore transfer. Strings of fertilized eggs are released by the females.

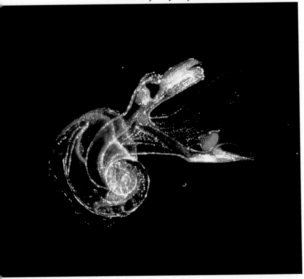

Family Atlantidae
125. *Atlanta* spp.
Identification: Dextrally coiled, smooth, flattened shell (diameter up to 4 mm) with a stabilizing keel on the largest whorl and a thin, transparent oval operculum. Relatively long and slender ventral fin has a distinct sucker. Body transparent, although the base of the keel may have an orange-brown to red-brown color, while the spire may be yellowish, brown or violet, often with a mottled pattern. *Natural History*: Atlantids are considered to be the most primitive heteropods. The body can be completely withdrawn into the shell when the animal is disturbed. Unlike other atlantids, both the shell and keel of *Atlanta* is calcified. The fin sucker holds veligers and small pteropod prey while the radula tears away pieces. Because of the shell, they are relatively inefficient swimmers. *Range and Habitat*: Found as far north as Vancouver Island. Depths to 150 meters.

Family Carinariidae
126. *Carinaria cristata*
Identification: Elongate body, up to 50 cm long but typically no more than 13 cm, with a relatively conspicuous external spiral shell covering the small visceral mass. Large muscular ventral fin with a single sucker directly opposite the visceral mass. May have reddish or blue pigmentation along margins of the ventral fin. Larger individuals often with tubercles on the surface of the body. *Natural History*: The most commonly encountered heteropod along central California. Rapid flexing of the body can be used for escape and changes of direction. Eats a wide variety of planktonic prey, with a preference for salps, doliolids and chaetognaths. Copepods and euphausiids may also be eaten. *Range and Habitat*: Temperate eastern and western North Pacific Ocean. Depths to 100 meters.

70

Family Pterotracheidae

127. *Firoloida desmaresti*

Identification: Elongate, cylindrical, transparent body (to 4 cm length); lacks a shell. Small visceral mass positioned posterior to the ventral swimming fin. Males possess a pair of tentacles posterior to the eyes and a sucker on the swimming fin. *Natural History*: Except for the black-pigmented eyes and food in the gut, this species is very difficult to see. During the night, individuals tend to hang relatively motionless with the ventral fin facing down while searching for bioluminescent prey, medusae, siphonophores, crustaceans and salps. Females retain a permanent egg filament from the posterior end. *Range and Habitat*: Widespread throughout oceanic areas of tropical and semitropical seas. Occasionally seen near surface off Baja California, and southern and central California.

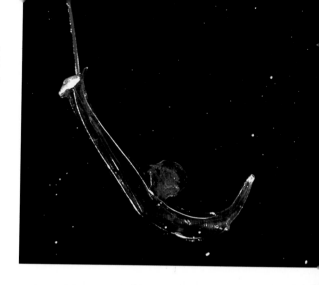

128. *Pterotrachea* spp.

Identification: Elongate, cylindrical, transparent body, up to 30 cm long; lacks a shell. The compact, opaque gut (situated posterior to the ventral swimming fin) and both eyes covered externally by a silvery reflective layer. Gelatinous "bib" on the ventral surface extends from the fin to the proboscis. Epidermis may be marked with pigment spots and the tail often has an elongate filament. Males possess a specialized sucker on the swimming fin. *Natural History*: Close approach produces rapid downward swimming involving undulation of the entire body. When not avoiding capture or pursuing prey, tends to remain motionless and curled up. Often associated with dense concentrations of siphonophores, which may be the preferred prey along with copepods and swimming polychaetes. *Range and Habitat*: Typically found in warm oceanic waters, but may range as far north as central California. *Remarks*: The gelatinous bib is wider on *P. scutata* as compared with the similar *P. coronata*.

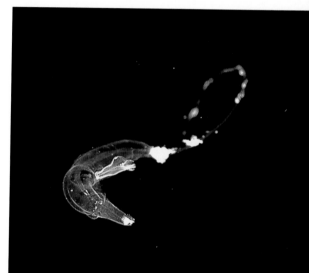

Subclass Opisthobranchia

The opisthobranchs are a diverse group of marine gastropods that includes the familiar nudibranchs and sea hares, in addition to the less well known pteropods. Within this subclass there is a tendency toward shell reduction or total elimination of the structure. The coiled body characteristic of most prosobranchs is largely absent. The mantle cavity may have one ctenidium or lack it completely, and many, such as the nudibranchs, possess surface gills.

Order Cephalaspidea
Family Gastropteridae

129. **Gastropteron pacificum**   Pacific wingfoot snail
Identification: Body like a shelled slug, to 33 mm long. Foot elongate with bilateral, flat, rounded extensions. Head, foot and lateral lobes with a yellow ochre color tinged with red, and with clusters of small purplish-red spots throughout. *Natural History*: This snail is primarily a benthic species that lives in soft sediments. It can swim by flapping the lateral foot lobes, but is not truly a pelagic animal. Swimming can be stimulated by contact with the predatory cephaspidian mollusc *Navanax* and may last several hours. Young individuals spend more time off the bottom and may be collected in nearshore plankton tows. *Range and Habitat*: Alaska (Aleutian Islands) to San Diego, California. To depths of at least 30 meters; swimming individuals can be found in the water column above their benthic habitats.

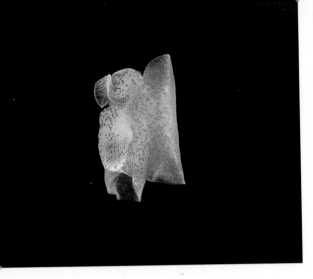

Order Thecosomata

The thecosomes are a group of about 50 species of planktonic opisthobranchs that along with gymnosomes are known as pteropods (for "wing foot"). All possess a mantle cavity (usually with a small gill) and a reduced foot with greatly enlarged epipodia used for swimming. Thecosomes feed passively while drifting by using a large external mucous web. The web and attached food particles are ingested by a mouth with lateral jaws and usually a radula. As indiscriminate omnivores, they trap a range of prey, including bacteria, copepods and other crustaceans, gastropod larvae, dinoflagellates, diatoms and radiolarians. Three families have calcareous shells, including all euthecosomes. Among pseudothecosomes, only the Peraclididae possess a shell, and are considered the most primitive (*Peraclis* is occasionally seen off Baja California). Thecosome adaptations for a planktonic existence include gelatinous bodies, thin shells, ciliary mucous feeding mechanism, protandrous hermaphroditic reproduction (starting as males) and production of floating egg masses. Due to the characteristic pattern of wing flapping, thecosome pteropods are commonly known as "sea butterflies."

Suborder Euthecosomata
Family Limacinidae

130. *Limacina helicina*
Identification: Body protected by a conspicuous, sinistrally coiled shell lacking a keel and with little or no ornamentation. Shell diameter up to 12–15 mm, but typically 1–3 mm off California (tend to be larger in northern part of range). Two wings are separate and lie dorsally and laterally to the foot lobes. While the body may have dark pigmentation, the shell surface is usually unpigmented. *Natural History*: Neutral buoyancy achieved with the large gelatinous wings and mucous web (diameter up to 6 cm). Exhibits diurnal vertical migrations and usually is at depth during daylight hours. Predatory fish may acquire stained flesh and an unpleasant odor after eating it. *Range and Habitat*: Common in arctic and temperate waters. Rare south of Pt. Conception. Oceanic at depths to 100 meters. *Remarks*: At least 7 species of *Limacina* have been described.

Family Cavoliniidae

131. *Clio pyramidata*

Identification: Bilaterally symmetrical uncoiled shell, length up to 10 mm, has a straight median axis with a pyramidal, shield-like shape. Fins are notched and the middle lobe of the foot is broad. Lacks the complex mantle lobes, operculum and pseudoconch found in *Cavolinia*. Surface of the shell characteristically covered with colonies of a particular type of hydroid. Mucous web diameter to 5 cm. *Natural History*: *Clio* can remain neutrally buoyant for long periods of time, and at night may move into shallower water. It is known to be an intermediate host for parasitic copepods that eventually infest fish. Sometimes comes up in trawls with only the pyramidal shell intact. *Range and Habitat*: Washington to Baja California (more common in southern part of range). Usually oceanic to depths of 400 meters.

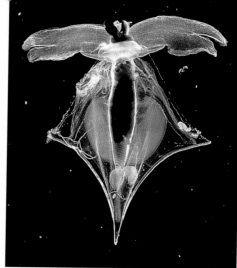

R. Gilmer

a. R. Gilmer b. G. Dietzmann

132. *Cavolinia tridentata*

Identification: The globose bilaterally symmetric shell, length up to 10 mm, is rounded and has three posterior projections. A pair of very distinctive, large ciliated mantle extensions pass through clefts in the shell. Mantle lobes completely enclose both the shell and the transparent gelatinous pseudoconch. Body and wings with a light brown color. *Natural History*: Feeds using a very large mucous web (diameter from 10 to 20 cm) which is wider than that formed by *C. inflexa* (which is typically 2 to 3 cm). Cavoliniid feeding webs are deflated and collected on the large platform of the foot lobes and then ingested as condensed strings. The shell may occasionally harbor a specific species of hydroid. The mantle appendages probably increase buoyancy and stability, and possess cilia to create water currents. *Range and Habitat*: Typically in warm oceanic waters, ranging as far north as southern California.

B. Upton (Monterey Bay Aquarium)

133. *Creseis virgula*

Identification: The pointed, dunce-cap shaped shell is bilaterally symmetrical, uncoiled and straight, with length up to 9 mm. The mantle lacks any extensions. *Natural History*: While feeding *Creseis* tends to sink slowly, although a gelatinous coating on part of the shell may help to reduce the sinking rate. An escape response consists of abandoning the web and sinking with the shell held vertical by keeping the wings motionless. The sharply pointed shell is known to cause skin irritation to swimmers who encounter large swarms. *Range and Habitat*: Cosmopolitan in tropical and semi-tropical waters. May be found as far north as central California. *Remarks*: *Creseis acicula*, which has a more thinly pointed shell, may occasionally be seen off Baja California.

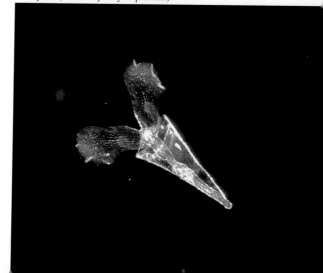

Suborder Pseudothecosomata
Family Cymbuliidae

134. *Corolla calceola*   Sea Butterfly

Identification: Transparent body lacks an external shell, with skeletal support and protection provided by an internal gelatinous pseudoconch. Slight indentations mark the margins of the oval wings which form a continous plate. Wingspan up to 8 cm. Relatively long proboscis partially fused with the wing plate; lacks a radula and jaws. A distinctive dark gut nucleus is present. *Natural History*: Slightly negatively buoyant, with position maintained by the wings along with the buoyant gelatinous tissue. Swimming is followed by a slow descending glide. Responds to disturbance by swimming rapidly. Forms a mucous sheet up to 2 meters in diameter. *Range and Habitat*: Temperate waters of the Pacific Ocean; more common north of Pt. Conception.

L. Bibko

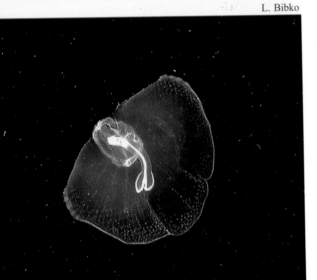

135. *Gleba cordata*

Identification: Flattened body (which distinguishes it from *Corolla*) lacks an external shell and has a thin transparent pseudoconch. Distinct indentations on the anterior margin divide each wing plate into two lobes. Wingspan up to 6 cm. Large prominent lateral wing plate mucous glands are more distinct than those of *Corolla*. Proboscis lacks a radula and jaws and is capable of great extension. May have brown or gold chromatophores on the swimming plate and melanophores on the pseudoconch. *Natural History*: Feeds in a manner similar to *Corolla*, with a mucous web up to 2 meters in diameter. This species is a strong, rapid swimmer. Studies have shown that it requires only about 1 to 2% of dry body weight/day for metabolic maintenance. *Range and Habitat*: Typically in warm oceanic waters, ranging north to central California.

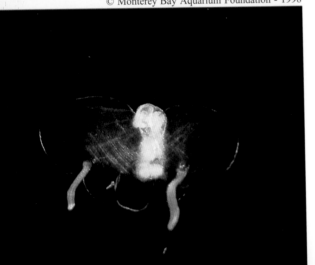

Family Desmopteridae

136. *Desmopterus papilio*

Identification: Barrel-shaped body, length from 2–3 mm, has a pair of wings which are fused to form a plate (span up to 6 mm). Lacks shell, pseudoconch and mantle cavity. Posterior margin of wing plate divided into five lobes, with each side possessing a fragile ciliated tentacle. Head bent ventrally, with a terminal mouth. Equipped with small jaws and a radula. The body may have a scattering of reddish-brown spots, particularly at the wing margins. *Natural History*: Typically hangs motionless, swimming in a loop pattern when disturbed. Unlike other thecosomes, lacks a gizzard and mucus producing glands. Because of its small size, this species is unlikely to be seen unless a careful search is made of the contents of a zooplankton tow. *Range and Habitat*: Cosmopolitan in tropical and subtropical oceanic waters.

Order Gymnosomata / Suborder Gymnosomata

Although also known as pteropods, gymnosomes (for "naked body") are probably not closely related to the thecosomes. About 50 species of these gastropods have been described. Because of the wings and general shape of the body, they are sometimes known as "sea angels." Gymnosomes lack a shell and mantle cavity, and are highly specialized predators. A pair of muscular swimming wings project from the sides of the body (derived from part of the foot). The wings are relatively small compared to thecosomes, but generally have a more rapid beat frequency. Ventrally between the wings is a remnant of the foot. The body is covered by a relatively tough elastic integument, which can contract, making identification of collected individuals difficult. A well-defined head with two pairs of tentacles is separated from the trunk by a narrow neck area. The mouth is located terminally on the head, with the buccal mass consisting of a radula, hook sacs (unique to gymnosomes) and a jaw. Lateral radula teeth and the chitinous hooks are used to grasp soft prey (primarily thecosomes) which are then pulled into the buccal cavity. Gymnosomes are simultaneous hermaphrodites, with a female genital pore situated on the right ventral side and the male copulatory organ located on the right side of the foot lobes.

Family Pneumodermatidae

137. *Crucibranchaea macrochira*

Identification: Body length from 8–14 mm with buccal apparatus withdrawn. A pair of thick feeding arms, each with 20–30 stalked suckers, and the buccal mass, are retracted when not feeding. The foot bears a pair of wide lateral lobes and a distinct long median lobe. At the posterior end of the body are gills with four branches. Body with scattered stellate chromatophores. *Natural History*: Symmetrical twisting action of the wings during upward and downward strokes enables relatively rapid swimming. Feeding tentacles and the buccal mass are extended by an increase in blood pressure through a slit-like opening that is not the true mouth. Various thecosome pteropods are the preferred prey. *Range and Habitat*: Cosmopolitan in tropical and semi-tropical waters of the Atlantic and Pacific Oceans. Occasionally ventures north to southern British Columbia.

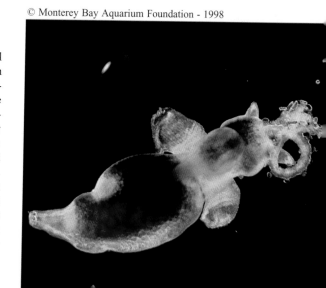

L. Madin

138. *Pneumodermopsis* sp.

Identification: The buccal apparatus contains a proboscis, with radula, hook sacs and jaw. Lateral feeding arms reduced compared to *Crucibranchaea*, with up to 16 suckers. The median arm has 3 to 5 suckers. Body usually with a dark blue or purple pigmentation. *Natural History*: This gymnosome is an agile swimmer that is among the swiftest pteropods (capable of swimming at 1 meter/second for short time periods). It feeds on various thecosome pteropods after grasping prey with lateral and median arm suckers. *Range*: May occasionally be found in oceanic waters off California and Baja California.

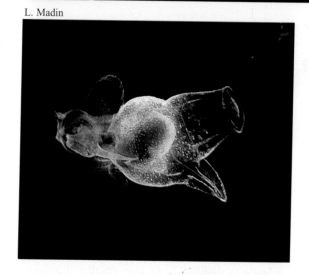

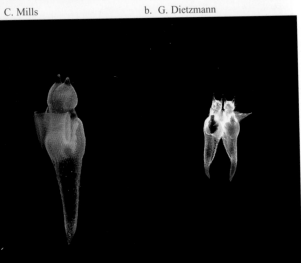

. C. Mills   b. G. Dietzmann

© Monterey Bay Aquarium Foundation - 1998

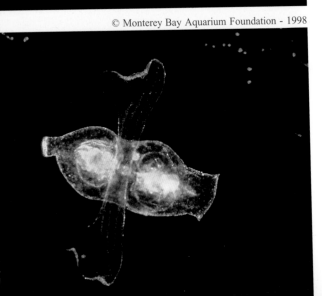

Family Cliopsidae

139. *Cliopsis krohni*   Sea Angel

Identification: Relatively large globular, flabby body, length to 4 cm, with a small visceral nucleus. Rounded swimming wings located very far forward near head. Integument may have a crinkled appearance in larger individuals, and generally lacks pigmentation. Smaller individuals possess a more elongate body having some color, with a brownish-gold cast to the body and rose tinge in the head area. *Natural History*: This species is among the slower swimming gymnosomes. Like other gymnosomes, *Cliopsis* is not known to regulate buoyancy with ionic balance or specialized buoyancy structures. Feeds on pseudothecosomes (typically *Corolla*). Upon contact with the prey feeding web, it ceases to swim while drawing the prey closer to the proboscis. *Range and Habitat:* Temperate oceanic waters north of southern California. To depths of 1500 meters; occasionally seen at the surface.

Family Clionidae

140. *Clione limacina*   Sea Angel

Identification: Body length up to 8 cm, typically less than 3 cm off west coast. Broad muscular swimming wings narrowly attached at sides of mid-line at anterior quarter of body. Three pairs of buccal cones (eversible tentacles without suckers) assist with prey capture. Lacks jaws, gills and external pigmentation. *Natural History*: Has a relatively fast, agile swimming motion. Its specialized diet consists solely of the thecosome, *Limacina*. Upon contact, hydraulic pressure extends the adhesive buccal cones to grasp the shell. Chitinous hooks and the radula pull the prey from the shell. Mating individuals unite ventrally for several hours of reciprocal fertilization. Encapsulated eggs are released in a free-floating spherical gelatinous mass. As with other gymnosomes, eggs hatch as veligers. *Range and Habitat:* Cosmopolitan in cold and temperate oceanic waters, to 600 meters.

141. *Thliptodon* sp.

Identification: Small cylindrical body, length to 25 mm, with loose-fitting skin. Head is relatively large, and a pair of paddle-like wings extend perpendicularly about mid-way along the body. The buccal apparatus possesses a radula and large hook sacs, but no jaw or other appendages. *Natural History*: Very little is known about this rare, inconspicuous pteropod, which can be found only by careful searching of plankton tows. It is among the slowest swimming gymnosomes. *Range and Habitat*: Deep water species that may occasionally be seen near the surface.

Order Nudibranchia

Nudibranchs are opisthobranch sea slugs that lack a shell and mantle cavity. Many species are colorful benthic dwellers, using a muscular foot to crawl slowly. A number of unrelated species are capable of swimming using lateral flexions of the entire body. Others use the dorsal cerata to row through the water. Swimming can be used to escape predators or search for new habitats, food or mates. Most are not nearly as adept at swimming as the pteropods and heteropods. In addition, a few nudibranchs have taken up a totally planktonic lifestyle, completing their entire life cycles in open water. All of these species feed to some extent on pelagic cnidarians. Reproduction and development are generally similar to benthic forms.

Suborder Dendronotacea
Family Dendronotidae

142. *Dendronotus iris*

Identification: Massive body, length up to 30 cm, with three to eight pairs of highly branched dorsal processes. A white line passes along the foot margin. Most have salmon-red bodies bearing orange, white-tipped processes, less commonly with light milky-purple bodies bearing light-orange processes tipped with white. *Natural History*: This distinctive nudibranch has large jaws used in clipping tentacles of the tube-dwelling anemone, *Pachycerianthus fimbriatus*. When threatened by predators such as the sunflower star, *D. iris* can swim by vigorous gyrations of the body. Although not an agile swimmer, this ability does provide an effective escape mechanism. Swimming is aided by a body that is more gelatinous than other nudibranchs. *Range and Habitat*: Aleutian Islands (Alaska) to Los Coronados (Baja California). Depths to 200 meters, sometimes at surface over deep water.

Family Phylliroidae

143. *Phylliroe atlantica*

Identification: Elongate, laterally compressed transparent body, length up to 5 cm, has a reduced foot and terminates with a slightly forked, expanded tail. Rhinophores on head are only body projections on the smooth, streamlined body. Forward from rhinophores is the snout bearing a terminal mouth. *Natural History*: This species is highly specialized for a planktonic existence. It swims using undulating movements of the entire body at speeds up to 15 cm/second. It feeds on hydromedusae and larvaceans. Young individuals enter hydromedusae such as *Zanclea* and slowly consume the steadily shrinking host. *Range and Habitat*: Tropical and subtropical Atlantic and Pacific. Occasionally to southern California waters. Epipelagic.

Family Tethyidae

144. ***Melibe leonina***   Hooded nudibranch

Identification: Gelatinous body up to 10 cm, with a distinctive oral hood, which bears tentacle-like extensions on the margin. Five or six pairs of leaf-like cerata extend from the dorsal surface. The body is typically yellowish brown to olive-green and pale gray. *Natural History*: This bizzare nudibranch is hard to confuse with any other. The narrow foot remains firmly attached to giant kelp or other substrates while feeding. The hood is swept downward, with the sides folding inward to trap zooplankton prey. Can form massive aggregations on a single giant kelp plant. In the Pacific northwest, *Melibe* typically occurs on eelgrass. Can swim clumsily by flexing the body from side to side. Swimming individuals can be common in surface waters following fall and winter storms when eelgrass beds break up. *Range and Habitat*: Alaska to Gulf of California.

L. Madin

Suborder Aeolidacea
Family Glaucidae

145. ***Glaucus atlanticus***

Identification: Flattened elongate body, length to 4 cm, has a short blunt head. Lobes on sides of body form three or four clusters of flattened papillae (cerata). The slender ventral foot ends in a long tail. Individuals exhibit countershading, with a deep blue-purple ventral surface and silvery white dorsal (lower) surface. *Natural History*: This nudibranch floats upside down at the surface and relies mostly on passive transport by wind and currents. Swallowed air is stored in the gastric cavity to maintain buoyancy. It preys on surface-dwelling cnidarians (*Velella*, *Porpita* and *Physalia*). Nematocysts from the prey become incorporated within the cerata as part of the defensive mechanism. *Range and Habitat*: Circumtropical. Occasionally ventures into southern California waters. Oceanic at the surface.

Family Fionidae

146. ***Fiona pinnata***

Identification: Typical eolid body does not differ to any great extent from benthic nudibranchs. Body length up to 6 cm, typically between 1.5–3.5 cm. Numerous sail-shaped cerata extend from sides of the dorsal surface. Depending on diet, color varies from dorsal blue-purple or brown to ventral white or pale yellow. *Natural History*: With little adaptation to a pelagic lifestyle, this species is not capable of swimming and lacks any buoyancy mechanism. Usually found clinging to prey (*Velella*, *Janthina* and pelagic barnacles), driftwood or floating seaweeds. Those that feed on *Velella* develop a deep blue coloration, while those feeding on barnacles assume a brown dorsal color. Growth to maturity is very rapid and the entire life cycle can be completed in one month. *Range and Habitat*: Circumtropical distribution.

Phylum Chordata / Subphylum Urochordata or Tunicata
Class Thaliacea

Thaliaceans are an ecologically important group of tunicates that have abandoned the sessile life style of benthic sea squirts in favor of a pelagic existence. Adults lack the tail typical of larval tunicates and have oral and atrial openings at opposite sides of the body. They do, however, retain the relatively rigid tunic. Water currents produced by the action of muscular loop-like bands in the transparent test serve for locomotion in salps and doliolids. The muscle bands are readily visible and among the most useful characteristics for distinguishing species of salps. Pyrosomes retain the more typical tunicate pattern of using cilia to move water through the body. All thaliaceans exhibit some form of alternation of sexual and asexual generations. Oozooids, formed from fertilized eggs, lack the capability of sexual reproduction, and instead produce asexual buds.

Order Salpida / Family Salpidae

The pelagic tunicates known as salps form a group of about 24 species of primarily tropical and subtropical gelatinous animals. They are among the fastest growing of all multicellular organisms and can play a significant role in marine ecosystems. Rhythmic contractions of circular muscles move water into the oral opening and out the atrial aperture, passing through the pharynx and atrial chamber. Movement by jet propulsion in either direction is accomplished by regulating the opening and closing of the anterior and posterior openings. Relaxation of body muscles allows the elastic tunic to expand, permitting water to enter orally as the anterior valves open. Swimming is aided by near-neutral buoyancy, which is achieved by exclusion of sulfate ions from the test. Salps are relatively energy efficient in moving water since they expel a large volume at low velocity. Pumping water also provides chemosensory information, food, gas exchange, waste removal and dispersal of sperm.

Feeding in salps is a continuous process closely linked to swimming. Phytoplankton and other particles as small as 1 micrometer are filtered, non-selectively. Like other tunicates, salps use a continuously renewed cone-shaped mucous net, which is secreted by the endostyle and supported by a pair of gill slits in the pharyngeal region. Cilia move the net into the esophagus for transport to the gut. When disturbed, salps often release the mucous net and cease to feed.

Populations of salps may form massive swarms of millions of individuals with incredible rapidity (population growth rates 10 to 100 times that of other zooplankton have been recorded). With their copious production of rapidly sinking fecal pellets, salps can serve to transfer significant amounts of food materials to organisms in mid-water and deep-sea benthic communities. Predators include heteropods, medusae, siphonophores, ctenophores, sea turtles, marine birds and many types of fish (at least 50 known species). With a dry-to-wet weight ratio typically of about 8%, salps provide a more concentrated food source compared to less dense ctenophores and cnidarian jellies. Salps also serve as traveling homes for certain copepods, hyperiid amphipods and several species of fish, which use their host for food, substrate, brood space and transport.

Salps have a complex life cycle consisting of alternating aggregate (sexually reproducing) and solitary (asexually reproducing) generations. Aggregate chains are produced by solitary asexual oozooids and may consist of tens to hundreds of individual blastozooids. Salps are protogynous hermaphrodites, with aggregates generally starting as females, whose eggs are fertilized by older male zooids. The embryo is attached to the body wall of the parent and develops directly into the oozooid. Rapid growth to maturity is the primary means for avoiding predation, since body defenses and sensory structures are not well developed.

Many salps can be difficult to identify, even by specialists, particularly when only a photograph is at hand. The number of muscle bands, connections between or among bands, and whether they are interrupted ventrally and dorsally are key means for distinguishing species. Even for the same species, various authors have counted the numbers of muscle bands differently, and citations may have slightly different numbers of muscle bands. It can be particularly confusing to count bands around the oral region where they tend to branch. The muscle band numbers described in this book should only be used as a general starting point for identification, and you may find they don't always agree with your observations. For most of the salps, we have included photographs of both the solitary and aggregate.

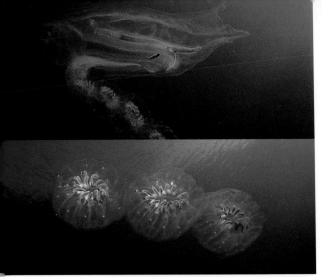

147. *Cyclosalpa affinis*

Identification: Solitary individuals with a relatively thick test, length up to 8 cm, with 7 ventrally interrupted muscle bands. The stolon, when present, projects through the test near band 2. Aggregate individuals have a thick soft test, length up to 8 cm, with 4 muscle bands. Form radial whorls, which may be united to form a chain of 3 or more circles. *Natural History*: *Cyclosalpa* are easy to distinguish from other salps when in the aggregate phase since no others form chains of linked whorls. As solitaries, the extended tubular gut (rather than a tight ball) is a good way to distinguish the cyclosalp group (which also includes *Helicosalpa*). As with most salps, pulsation rate tends to decline as water temperature drops. Can filter food particles of 4 micrometers or larger with 100% efficiency. *Range and Habitat*: Tropical and temperate oceanic waters. Northern range throughout California and occasionally to Gulf of Alaska.

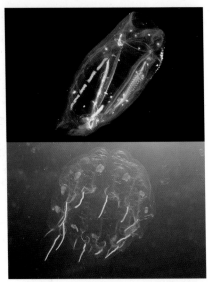

148. *Cyclosalpa bakeri*

Identification: Solitary individuals pear-shaped with the posterior end smaller, length from 5–15 cm. Delicate, slow swimmer with a thin, flabby test and 7 dorsally and ventrally interrupted body muscles. Pairs of distinct lateral whitish patches lie between each. When present, the stolon emerges from the test at band 2. Aggregate individuals have a thin flabby test, and form radial whorls up to 20 cm in diameter, with about a dozen zooids. Each individual has a pair of posterior end projections and 4 body muscles. The gut forms a loop at the base of the projections. *Natural History*: Tends to rise to the surface at night, which probably serves to concentrate the population prior to sperm release for fertilizing eggs. An individual can consume over half its body mass in 24 hours. Aggregates can increase body size by about 25% per day. *Range and Habitat*: Tropical and temperate oceanic waters. Northern range to Gulf of Alaska. More common than *C. affinis*.

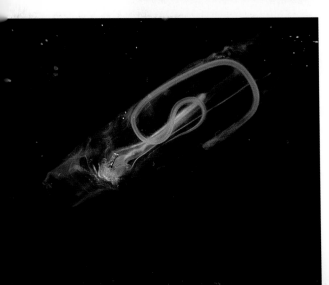

149. *Helicosalpa virgula*

Identification: Solitary with large body, up to 15 cm length. Test thick, flabby and relatively gelatinous with 8 muscle bands. Can be distinguished from *Cyclosalpa* by the presence of a ventral longitudinal muscle. Lateral white patches (sometimes called "light organs") form a line ranging from muscle band 1 to 6. The stolon, when present, is very obvious and does not produce a whorl. Aggregate individuals have a firm globular test, length up to 3.5 cm, with a posterior projection which is directed dorsally and forward. Muscle tissue in the test does not form distinct bands. Zooids are asymmetrical and alternate right and left hand forms along the length of the chain. *Range and Habitat*: Cosmopolitan; known from the Mediterranean, mid-Atlantic, central Indian Ocean and eastern Pacific. Seen rarely off California.

150. *Iasis zonaria*

Identification: Solitary individuals have a very firm, elongate test, length from 1.5–5 cm. The test possesses three posterior dorsally projections and 6 distinctively broad dorsally and ventrally interrupted muscle bands (bands wider than spaces between them). Mouth forms a broad slit, while the gut nucleus resides as a small mass near the rear of the body. Aggregate individuals have a firm asymmetric test, length from 2 to 6 cm, with a rear projection and 5 broad ventrally interrupted body muscles. The branchial openings are directed dorsally. *Natural History*: As with *Salpa*, the swimming thrust of the chain is aligned with the axis, which enables more efficient swimming. Among the fastest swimming and most cold tolerant salps. *Range and Habitat*: Relatively abundant in Atlantic, Pacific to Gulf of Alaska and Indian Oceans. Oceanic to 200 meters.

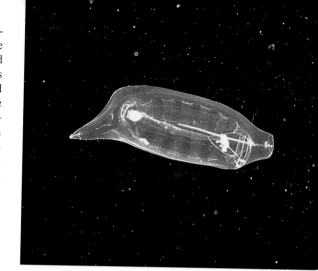

a. G. Dietzmann

151. *Pegea confoederata*

Identification: Solitary individuals with a plump body, length to 14 cm. Four body muscles combine dorsally to form a distinct pair of X's. Gut forms a tight nucleus. Body with a reddish-brown cast throughout. Stolon makes two complete turns around the gut with a chain of 150 to 200 zooids. Aggregate individuals have a cylindrical body with floppy test and length from 6–12 cm. Forms a double-row chain coiled into a tight spiral plane, with the axes of the zooids at right angles to the chain axis. *Natural History*: When given a sufficiently dense concentration of phytoplankton, feeding is disrupted, which may prevent this species from forming massive blooms. Swims slower than various species of *Salpa*. *Range and Habitat*: Moderately abundant in tropical and subtropical oceanic waters, to depths of 150 meters. Northern range to central California. *Remarks*: *P. socia*, with gold-colored pigmentation, is more likely to be seen north of central California.

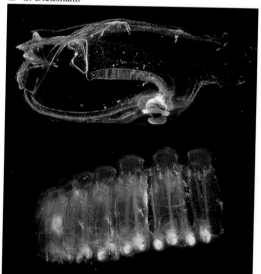

b. D. Wrobel

152. *Salpa fusiformis*

Identification: Solitary individuals with test length from 1 to 5 cm, with 9 body muscles. Gut forms a tight nucleus. Aggregate individuals have a smooth fusiform-shaped test, length from 0.5–4 cm, with one anterior and one posterior end projection and 6 body muscles. Forms linear chains with the zooid axes aligned along the chain axis. Aggregate zooids are often seen singly. *Natural History*: Unlike most salps, this species can undergo daily vertical migrations of up to 500 meters, occurring near the surface primarily at night. Aggregate chains move with surprising speed and may have a snake-like appearance. Maximum individual growth rates of about 40% length increase per day have been measured. *Range and Habitat*: Atlantic, Pacific and Indian Oceans. Northern range to Bering Sea. Primarily oceanic with occasional nearshore swarms.

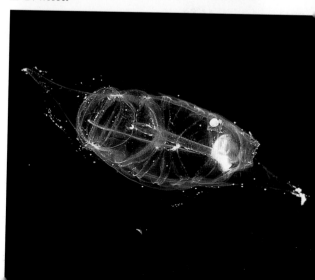

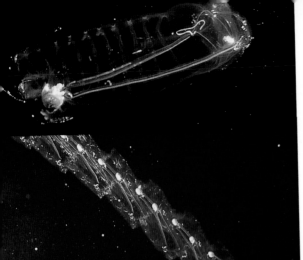

153. *Salpa maxima*

Identification: Solitary individuals with a thick, firm test, length up to 23 cm, and 8 body muscles. Posterior with numerous spines. Gut forms a tight nucleus. Aggregate individuals have a thick, smooth test, length up to 12 cm, with 6 muscles and end projections. Form linear chains with the zooid axes aligned along the chain axis. Can be difficult to distinguish from *S. fusiformis*, other than the much larger potential size. *Natural History*: Species of *Salpa* have higher swimming speeds (up to 6 cm/sec) and respiration rates than other salps. Parallel axis and thrust alignment of individuals in aggregates helps to streamline chains for more efficient swimming. With their large body size *S. maxima* can filter up to 2.5 liters of water per hour. *Range and Habitat*: Relatively rare in surface waters of Atlantic and Pacific Oceans, with a northern range to Bering Sea.

L. Madin

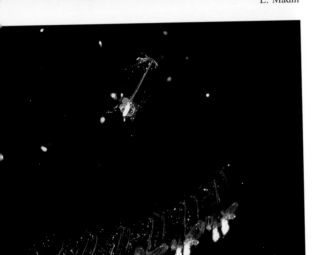

154. *Thalia democratica*

Identification: Solitary individuals with a thick test, length up to 12 mm, bearing a pair of lateral posterior projections and 5 body muscles. Aggregate individuals have a thick, stiff asymmetrical test, length up to 6 mm, bearing prominent ridges and grooves, and 5 muscle bands. There can be a great amount of variation in test structure with some being smooth and flabby. Smallest of the salps along the west coast. *Natural History*: Growth rates among the fastest of any multicellular animal, increasing in length 10-20% an hour. Generation times range from several weeks to only several days. This species is the most common central and southern California salp. Massive aggregations may form, with a swarm off southern California once observed that covered 3500 square miles with a total of billions of individuals. *Range and Habitat*: Widespread and common in tropical and temperate oceanic waters, usually near the surface.

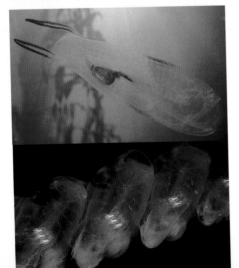

155. *Thetys vagina*

Identification: Solitary individuals with a firm, thick spiny test, length up to 33 cm, with about 20 weakly developed partial muscle bands. A distinctive pair of pigmented posterior projections are useful for identification. The large, compact gut is easily visible. Aggregate individuals have a thick, firm asymmetric test, length up to 25 cm, with a cylindrical body and 5 muscle bands (dorsally only). The body surface may have broad scattered processes. Oral and atrial openings are turned to one side. Form double-row chains with zooid axes not quite at right angles to the chain. *Natural History*: Oozooids of this species are the largest salp along the west coast. The solid test imparts a distinct rigidity when touched, and retains its shape when taken out of the water. *Range and Habitat*: Atlantic, Pacific and Indian Oceans, at depths from surface to 150 meters.

156. *Weelia cylindrica*

Identification: Solitary individuals with a uniform cylindrical body, thin soft test with 9 muscle bands and terminal oral and atrial openings. Small spherical gut nucleus near the posterior end. Aggregates form linear chains with the zooid axes aligned along the chain axis. Each aggregate individual has 5 body muscles, with oral and atrial openings directed dorsally. *Natural History*: This species is relatively rare throughout the west coast range. It is closely related to the various species of *Salpa*, and is difficult to distinguish from them without close examination. *Range and Habitat*: Atlantic and Pacific Oceans. Northern range to Bering Sea. Epipelagic. *Remarks*: This salp is sometimes placed in the genus *Salpa*.

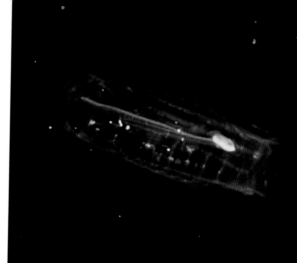

L. Madin

Order Doliolida

The doliolids are a group of inconspicuous, transparent pelagic tunicates inhabiting continental shelf waters of tropical and semi-tropical seas. Unlike salps, which must continuously swim to feed, doliolids pump water through their mucous feeding net using gill cilia. This limits filtration rates to values much less than that of salps, but allows them to feed while motionless. Water passes from the pharyngeal cavity through the atrial aperture, leaving food particles trapped within a mucous filter secreted by the endostyle. Doliolids feed on small particles, including bacteria and phytoplankton. The mucus is wrapped into a cord at the dorsal midline of the pharyngeal cavity and drawn into the mouth for ingestion. Species identification is difficult, and is based on the position of the internal organs relative to the muscle bands.

Doliolids have a characteristic jumpy swimming pattern when disturbed. The escape response consists of jet propulsion by rapid contraction of the circumferential muscles and the animal may cover many body lengths with each pulse. Directional control is possible by closure of either the atrial or branchial aperture. Expansion of the elastic test brings the body to its resting volume following muscle relaxation. The normal pattern is to slowly sink, alternating with single contractions at irregular intervals which drive the doliolid obliquely upwards.

The life cycle of doliolids is more complex than that of other thaliaceans. The oozooid develops reduced internal organs but retains a strong swimming capability. It produces a dorsally positioned "tail" of asexual embryonic zooids (forming the cadophore), with lateral rows of nutritive gastrozooids and a median row of phorozooids. At this stage the oozooid is known as a nurse. Other median row zooids develop into gonozooids which attach to the phorozooids, and are then carried away together. These form sexually mature gonozooids and separate from the phorozooids. The solitary gonozooids are hermaphroditic and produce the tadpole-shaped larvae that develop into oozooids. Of all the stages, the gonozooids are typically the most abundant since they are produced asexually in chains. With the high feeding rates and asexual reproduction characteristic of doliolids, rapid population increases and resultant depletion of planktonic food within an area are possible.

D. Wrobel   b. L. Madin

Family Doliolidae

157. *Dolioletta gegenbauri*

Identification: Barrel-shaped body with 8 distinct circumferential muscle bands (oozooid with 9 bands). Gonozooid body length up to 9 mm. The gills have a ventral attachment at band 5 and an anterior dorsal attachment at band 3. Anterior aperture typically with 12 lobes, while the posterior has 8. *Natural History*: This species is the most cold tolerant doliolid. It is capable of feeding on particles ranging from less than 5 micrometers to greater than 100 micrometers with equal efficiency. Concentrated swarms are capable of clearing the phytoplankton where they are living in less than a day. A single oozooid can lead to the production of thousands of gonozooids in several days. Can be very abundant over wide ranges, with densities up to 5 to 500 per cubic meter, and may be the most abundant thaliacean off California. *Range and Habitat*: Oceanic off California, with occasional nearshore swarms. More common north of Pt. Conception, preferring water cooler than 15°C.

Order Pyrosomatida

Pyrosomes are a group of largely tropical colonial thaliaceans. Some species form colonies that may approach 4 meters in length, but colonies only several cm long are most common. They feed in a manner similar to doliolids by using cilia to move water past a mucous filter. Of the thaliaceans, pyrosomes are closest in structure and function to benthic compound ascidians since they depend more on cilia to move water through the body. Colonies are formed by zooids arranged with the oral apertures on the outside, with water passing through these openings into the center of the colonial tube. This action results in motion of the colony with the open end trailing. Pyrosomes are brightly bioluminescent at times - it has been estimated that the light emitted might be visible in clear water for 100 meters. Such flashes have been startling and delighting sailors for millennia.

Family Pyrosomatidae

158. *Pyrosoma atlanticum*

Identification: Tightly packed individuals embedded in a tough, rigid tube, closed and slightly tapered at one end. Zooids with length up to 8.5 mm; colony length to 60 cm. Open end terminates with a diaphragm. Tests with prominent papillae. Tube may be colorless, pink, or grayish to blue-green, with the colonial wall opaque. *Natural History*: The most widespread and common pyrosome in middle and low latitudes. Produces intense blue-green bioluminescene. Zooids use flashes of light to signal adjacent zooids to cease ciliary motion when disturbed. High feeding rates result in the copious production of fecal pellets. This species is a vertical migrator, with a range of 750 meters, and occasionally forms dense surface swarms. *Range and Habitat*: Tropical and temperate Atlantic, Pacific and Indian Oceans.

Class Larvacea / Appendicularia

The appendicularians (also known as larvaceans) are unique among the tunicates in the retention of the larval "tadpole" form as sexually mature adults. The tail, which is lacking in thaliaceans, retains a notochord and is used both for locomotion and to produce feeding currents. It typically connects to the body ventral to the stomach. Gill slits form an opening from the pharynx to the exterior. Adults lack the tunic common to other tunicates, but use a complicated external mucous net structure (the "house") produced by a gland known as the oikoplast for collecting planktonic food. Shallow water species are generally very small and difficult to see, but may be present near the surface in very high densities. New houses are produced when the old ones clog with food, so both occupied and unoccupied houses occur together in the water column, the clogged older mucous houses being more visible. With the exception of *Oikopleura dioica*, larvaceans are hermaphrodites.

Order Copelata
Family Oikopleuridae

159. *Bathochordaeus charon*

Identification: Body much larger than other larvaceans, with a length up to 25 mm. Notochord clearly visible along the central axis of the relatively short, broad trail. The only conspicuous internal structures are the gut and gonads (when mature). *Natural History*: Secretes a continuous sheet of mucus (up to 2 meters diameter) that surrounds the feeding filter and body. It protects the animal and strains large particles from water passing to the inner filter. Particle-laden mucous sheets are important contributors to transport of organic material to deeper habitats. The discarded houses support entire communities of detritus-feeding zooplankton and various microbes. The tail beats slowly to pump water and food particles into the mid-line of the feeding filter. *Range and Habitat*: Cosmopolitan in midwater between 35°N and 35°S latitude, at depths from 100 to 500 meters.

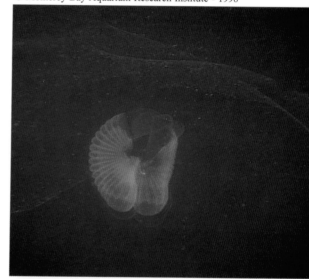

P. Flood

160. *Oikopleura* spp.

Identification: Oval body with tail, having a tadpole-like appearance. Body length 1 mm (*O. dioica* and *O. longicauda*); 2.0 to 2.5 mm (*O. labradoriensis*). Tail typically 3 to 4 times the length of the body. Individuals are usually surrounded by a mucous house that ranges from about the size of a pea to the size of a walnut. *Natural History*: The mucous house functions for feeding, protection from predators and buoyancy. Food particles are concentrated and transferred to the adjacent mouth. From 5 to 10 houses may be produced and discarded in one day. Houses are capable of producing endogenous blue-green luminescent flashes that make a significant contribution to surface bioluminescence. *Range and Habitat*: Cosmopolitan throughout tropical and temperate seas. *Oikopleura vanhoeffeni* (boreal and arctic waters) and *O. labradoriensis* are more cold tolerant than *O. dioica*. Usually no deeper than 200 meters. *Remarks*: The cosmopolitan species *Fritillaria borealis* is distinguished by the connection of the tail at the midpoint of the body.

Selected References

Cnidarians

Alvariño, A. 1971. Siphonophores of the Pacific, with a review of the world distribution. Bulletin of the Scripps Institution of Oceanography, 16:1-432.

Arai, M.N. 1997. A Functional Biology of Scyphozoa. Chapman and Hall, London, 316 pp.

Arai, M.N. and A. Brinckmann-Voss. 1980. Hydromedusae of British Columbia and Puget Sound. Canadian Bulletin of Fisheries and Aquatic Sciences. 204, 192 pp.

Bigelow, H.B. 1911. Reports on the scientific results of the expedition to the eastern tropical Pacific by the U.S. Fish Commission steamer "Albatross" 1904-1905. XXII. The siphonophorae. Memoirs of the Museum of Comparative Zoology, Harvard College, 38:171-401, 32 pls.

_____ 1912. Reports on the scientific results of the expedition to the eastern tropical Pacific by the U.S. Fish Commission steamer "Albatross" 1904-1905. XXVI. The ctenophores. Bulletin of the Museum of Comparative Zoology, Harvard College, 54:369-404, 2 pls.

Bouillon, J., F. Boero, F. Cicogna, J.M. Gili and R.G. Hughes. 1992. Non-siphonophoran Hydrozoa: what are we talking about? Scientia Marina, 56:279-284.

Brinckmann-Voss, A. 1985. Hydroids and medusae of *Sarsia apicula* (Murbach and Shearer, 1902) and *Sarsia princeps* (Haeckel, 1879) from British Columbia and Puget Sound with an evaluation of their systematic characters. Canadian Journal of Zoology, 63:673-681.

Costello, J.H. and S.P. Colin. 1995. Flow and feeding by swimming scyphomedusae. Marine Biology, 124:399-406.

Foerster, R.E. 1923. The Hydromedusae of the west coast of North America, with special reference to those of the Vancouver Island Region. Contributions to Canadian Biology (New Series), 1:219-277, 5 pls.

Hand, C. 1954. Three Pacific species of "Lar" (including a new species), their hosts, medusae, and relationships (Coelenterata, Hydrozoa). Pacific Science, 8:51-67.

Kirkpatrick, P.A. and P.R. Pugh. 1984. Siphonophores and Velellids. E.J. Brill / Dr. W. Backhuys, London, 154 pp.

Kramp, P.L. 1961. Synopsis of the medusae of the world. Journal of the Marine Biological Association, U.K., 40:1-469.

_____. 1968. The Hydromedusae of the Pacific and Indian Oceans: Sections II and III. "Dana" Report, 72:1-200.

Larson, R.J. 1976. Cubomedusae: feeding - functional morphology, behavior and phylogenetic position. Pp. 237-245 *In* Coelenterate Ecology and Behavior, G.O. Mackie (ed.). Plenum Press, New York.

_____. 1979. Feeding in coronate medusae (Class Scyphozoa, Order Coronatae). Marine Behavior and Physiology, 6:123-129.

_____. 1980. The medusa of *Velella velella* (Linnaeus, 1758) (Hydrozoa, Chondrophorae). Journal of Plankton Research, 2:183-186.

_____. 1990. Scyphomedusae and Cubomedusae from the eastern Pacific. Bulletin of Marine Science, 47:546-556.

_____. 1992. Riding Langmuir circulations and swimming in circles: a novel clustering behavior by the scyphomedusa *Linuche unguiculata*. Marine Biology, 112:229-235.

Larson, R.J., and A.C. Arneson, 1990. Two medusae new to the coast of California: *Carybdea marsupialis* (Linnaeus, 1758), a Cubomedusa and *Phyllorhiza punctata* von Lendenfeld, 1884, a rhizostome Scyphomedusa. Bulletin of the Southern California Academy of Science, 89, 130-136.

Larson, R.J., and D.G. Fautin, 1989. Stauromedusae of the genus *Manania* (=*Thaumatoscyphus*) (Cnidaria, Scyphozoa) in the northeast Pacific, including descriptions of new species *Manania gwilliami* and *Manania handi*. Canadian Journal of Zoology, 67:1543-1549.

Mackie, G.O. and G.V. Mackie. 1963. Systematic and biological notes on living hydromedusae from Puget Sound. Contributions in Zoology, National Museum of Canada, Bulletin, 199:63-84.

Mackie, G.O., P.R. Pugh, and J.E. Purcell. 1987. Siphonophore biology. Advances in Marine Biology, 24:97-262.

Mills, C.E., 1981. Seasonal occurrence of planktonic medusae and ctenophores in the San Juan Archipelago (NE Pacific). Wasmann Journal of Biology, 39:6-29.

_____. 1981. Diversity of swimming behaviors in hydromedusae as related to feeding and utilization of space. Marine Biology, 64: 185-189.

_____. 1987. *In situ* and shipboard studies of living hydromedusae and hydroids: preliminary observations of life cycle adaptations to the open ocean. Pp. 197-207 *In* Modern Trends in the Systematics, Ecology and Evolution of Hydroids and Hydromedusae. J. Bouillon, F. Boero, F. Cicogna and P.F.S. Cornelius (eds.). Clarendon Press, Oxford.

_____. 1996. Keys to the Hydrozoan Medusae (pp. 32-44, 487-489), Siphonophora (pp. 62-65, 489), Scyphozoa: Semaeostomae (pp. 65-67) and Stauromedusae (p 489). *In* Marine Invertebrates of the Pacific Northwest. E.N. Kozloff (ed.). University of Washington Press, Seattle.

Mills, C.E., and F. Sommer. 1995. Invertebrate introductions in marine habitats: two species of hydromedusae (Cnidaria) native to the Black Sea, *Maeotias inexspectata* and *Blackfordia virginica*, invade San Francisco Bay. Marine Biology, 122:279-288.

Naumov, D.V. 1969. Hydroids and hydromedusae of the USSR (J. Salkind, Israel Program for Scientific Translations, Trans.). Israel Program for Scientific Translations, Jerusalem, 660 pp.

Purcell, J.E. 1981. Dietary composition and diel feeding patterns of epipelagic siphonophores. Marine Biology, 65:83-90.

Rees, J.T. and R.J. Larson. 1980. Morphological variation in the hydromedusa genus *Polyorchis* on the west coast of North America. Canadian Journal of Zoology, 58, 2089-2095.

Russell, F.S. 1953. The Medusae of the British Isles. Cambridge University Press, Cambridge, 530 pp., 35 pls.

_____. 1970. The Medusae of the British Isles. II. Pelagic Scyphozoa. Cambridge University Press, Cambridge, 284 pp.

Segura-Puertes, L. 1984. Morphology, systematics and zoogeography of medusae (Cnidaria: Hydrozoa and Scyphozoa) from the eastern tropical Pacific. Inst. Cienc. del Mar y Limnol. Univ. Nal. Auton. Mexico Publ. Esp., 8:1-320.

Skogsberg, T. 1948. A systematic study of the Family Polyorchidae (Hydromedusae). Proceedings of the California Academy of Sciences, Fourth Series, 26(5):101-124.

Strand, S.W. and W.M. Hamner. 1988. Predatory behavior of *Phacellophora camtschatica* and size-selective predation upon *Aurelia aurita* (Scyphozoa: Cnidaria) in Saanich Inlet, British Columbia. Marine Biology, 99:409-414.

Stretch, J.J., and J.M. King. 1980. Direct fission: an undescribed reproductive method in hydromedusae. Bulletin of Marine Science, 30:522-525.

Thuesen, E.V. and J.J. Childress. 1994. Oxygen consumption rates and metabolic enzyme activities of oceanic California medusae in relation to body size and habitat depth. Biological Bulletin, 187:84-98.

Torrey, H.B. 1909. The Leptomedusae of the San Diego Region. University of California Publications in Zoology, 6:11-31.

Totton, A.K. 1965. A Synopsis of the Siphonophora. Trustees of the British Museum (Natural History), London, 230 pp, 40 pls.

Uchida, T. 1929. Studies on the Stauromedusae and Cubomedusae, with special reference to their metamorphosis. Japanese Journal of Zoology, 2:103-193.

Ctenophores

Berkeley, C. 1930. Symbiosis of a *Beroe* and a flagellate. Contribution to Canadian Biology and Fisheries, New Series, 6:15-21.

Chun, C. 1880. Fauna und Flora des Golfes von Neapel. I. Ctenophorae. Wilhelm Engelmann, Leipzig, 313 pp, 16 pls.

Cormier, M.J., K. Hori, and J.M. Anderson. 1974. Bioluminescence in coelenterates. Biochimica Biophysica Acta, 346:137-164.

Easterly, C.O. 1914. A study of the occurrence and manner of distribution of the Ctenophora of the San Diego region. University of California Publications in Zoology, 13:21-38.

Gyllenberg, G. and W. Greve. 1979. Studies on oxygen uptake in ctenophores. Ann. Zool. Fenn, 16:44-49.

Haddock, S.H.D. and J.F. Case. 1995. Not all ctenophores are bioluminescent: *Pleurobrachia*. Biological Bulletin, 189:356-362.

Harbison, G.R. 1984. On the classification and evolution of the Ctenophora. Pp. 78-100 *In* The Origins and Relationships of Lower Invertebrates. S.C. Morris, J.D. George, R. Gibson and H.M. Platt (eds.). Oxford University Press, Oxford.

Harbison, G.R., D.C. Biggs and L. P. Madin. 1977. The associations of Amphipoda Hyperiidea with gelatinous zooplankton - II. Associations with Cnidaria, Ctenophora and Radiolaria. Deep-Sea Research, 24:465-488.

Harbison, G.R., L.P. Madin and N.R. Swanberg. 1978. On the natural history and distribution of oceanic ctenophores. Deep-Sea Research, 25:233-256.

Harbison, G.R. and L.P. Madin. 1982. The Ctenophora. Pp. 707-715 *In*: Synopsis and classification of living organisms, Vol. 1. S.P. Parker (ed.). McGraw Hill, New York.

Harbison, G.R. and R.L. Miller. 1986. Not all ctenophores are hermaphrodites. Studies on the systematics, distribution, sexuality and development of two species of *Ocyropsis*. Marine Biology, 90:413-424.

Hirota, J. 1974. Quantitative natural history of *Pleurobrachia bachei* in La Jolla Bight. Fishery Bulletin, 72:295-335.

Hoeger, U. 1983. Biochemical composition of ctenophores. Journal of Experimental Marine Biology and Ecology, 72:251-261.

Horridge, G.A. 1974. Recent studies on the Ctenophora. Pp. 439-468 *In*: Coelenterate Biology. L. Muscatine and H.M. Lenhoff (eds.). Academic Press, New York,.

Madin, L.P. 1988. Feeding behavior of tentaculate predators: *in situ* observations and a conceptual model. Bulletin of Marine Science, 43:413-429.

Matsumoto, G.I. 1990. Ctenophores of the Monterey Canyon (including a general key, photographs, and slides). Monterey Bay Aquarium Research Institute, Pacific Grove, California, 34 pp.

Matsumoto, G.I. and W.M. Hamner. 1988. Modes of water manipulation by the lobate ctenophore *Leucothea* sp. Marine Biology, 97:551-558.

Matsumoto, G.I. and G.R. Harbison. 1993. *In situ* observations of foraging, feeding, and escape behavior in three orders of oceanic ctenophores: Lobata, Cestida, and Beroida. Marine Biology, 117:279-287.

Mayer, A.G. 1912. Ctenophores of the Atlantic Coast of North America. Carnegie Institution of Washington, Publication No. 162, 58 pp.

Mills, C.E. 1996. Key to the phylum Ctenophora. Pp.79-81, 490-491 *In* Marine Invertebrates of the Pacific Northwest. E.N. Kozloff (ed.). University of Washington Press, Seattle.

Mills, C.E. and N. McLean. 1991. Ectoparasitism by a dinoflagellate (Dinoflagellata: Oodinidae) on 5 ctenophores (Ctenophora) and a hydromedusa (Cnidaria). Diseases of Aquatic Organisms, 10:211-216.

Mills, C.E. and R.L. Miller. 1984. Ingestion of a medusae (*Aegina citrea*) by the nematocyst-containing ctenophore *Haekelia rubra* (formerly *Euchlora rubra*): phylogenetic implications. Marine Biology, 78:215-221.

Moser, F. 1909. Die Ctenophoren der deutsche Südpolar-Expedition 1901-1903. Deutsche Südpolar-Expedition, XI. Zoologie III:116-192, 3 pls.

Pianka, H.D. 1974. Ctenophora. Pp.201-265 *In* Reproduction of Marine Invertebrates, Volume I: Acoelomate and Pseudocoelomate Metazoans. A.C. Giese and J.S. Pearse (eds.). Academic Press, New York.

Reeve, M.R. 1980. Comparative experimental studies on the feeding of chaetognaths and ctenophores. Journal of Plankton Research, 2:381-393.

Reeve, M.R. and M.A. Walter. 1978. Nutritional ecology of ctenophores - a review of recent research. Advances in Marine Biology, 15:249-287.

Stanford, L.O. 1931. Studies on ctenophores of Monterey Bay. MS Thesis, Stanford University, 73 pp.

Stoecker, D.K., P.G. Verity, A.E. Michaels and L.H. Davis. 1987. Feeding by larval and post-larval ctenophores on microzooplankton. Journal of Plankton Research, 9:667-683.

Stretch, J.J. 1982. Observations on the abundance and feeding behavior of the cestid ctenophore, *Velamen parallelum*. Bulletin of Marine Science, 32:796-799.

Tamm, S.L. 1980. Cilia and ctenophores. Oceanus, 23:50-59.

Tamm, S.L. and S. Tamm. 1993. Diversity of macrociliary size, tooth patterns, and distribution in *Beroe* (Ctenophora). Zoomorphology, 113:79-89.

_____. 1991. Reversible epithelial adhesion closes the mouth of *Beroe*, a carnivorous marine jelly. Biological Bulletin, 181:463-473.

Torrey, H.B. 1904. The ctenophores of the San Diego region. University California Publications in Zoology, 2:45-51.

Molluscs

Bé, A.W.H. and R.W. Gilmer. 1977. A zoogeographic and taxonomic review of euthecosomatous pteropoda. Pp. 733-808 *In*: Oceanic Micropaleontology, Vol. 1. A.T.S. Ramsey (ed.). Academic Press, London.

Bertsch, H. 1969. A note on the range of *Gastropteron pacificum*. Veliger, 11:431-433.

Dales, R.P. 1952. The distribution of some heteropod molluscs off the Pacific coast of North America. Proceedings of the Zoological Society of London, 122:1007-1015.

Davenport, J. and A. Bebbington. 1990. Observations on the swimming and buoyancy of some thecosomatous pteropod gastropods. Journal of Molluscan Studies, 56:487-497.

Farmer, W.M. 1970. Swimming gastropods (Opisthobranchia and Prosobranchia). Veliger, 13:73-89.

Gilmer, R.W. 1972. Free-floating mucus webs: a novel feeding adaptation for the open ocean. Science, 176:1239-1240.

_____. 1990. *In situ* observations of feeding behavior of thecosome pteropod molluscs. American Malacological Bulletin, 8:53-59.

Gilmer, R.W. and G.R. Harbison. 1986. Morphology and field behavior of pteropod molluscs: feeding methods in the families Cavoliniidae, Limacinidae and Peraclididae (Gastropoda: Thecosomata). Marine Biology, 91:47-57.

Lalli, C.M. 1970. Structure and function of the buccal apparatus of *Clione limacina* (Phipps) with a review of feeding in gymnosomatous pteropods. Journal of Experimental Marine Biology and Ecology, 4:101-118.

Lalli, C.M. and R.W. Gilmer. 1989. Pelagic Snails: The Biology of Holoplanktonic Gastropod Mollusks. Stanford University Press, Stanford, California, 259 pp.

Marcus, E. 1971. Range of *Gastropteron pacificum* Bergh, 1893. Veliger, 13:297.

McGowan, J.A. 1967. Distributional atlas of pelagic molluscs in the California Current region. CALCOFI Atlas No.6 (California Marine Research Committee, 218 pp.

_____. 1968. The Thecosomata and Gymnosomata of California. Veliger, 3 (Supplement):103-125.

Mills, C.E. 1994. Seasonal swimming of sexually mature benthic opisthobranch molluscs (*Melibe leonina* and *Gastropteron pacificum*) may augment population dispersal. Pp. 313-319 In Reproduction and Development of Marine Invertebrates. S.A. Stricker, W.H. Wilson Jr. and G.L. Shinn (eds.). Johns Hopkins University Press, Baltimore.

Seapy, R.R. 1985. The pelagic genus *Pterotrachea* (Gastropoda: Heteropoda) from Hawaiian waters: A taxonomic review. Malacologia, 26:125-135.

Seapy, R.R. and R.E. Young. 1986. Concealment in epipelagic pterotracheid heteropods (Gastropoda) and cranchiid squids (Cephalopoda). Journal of the Zoological Society of London (A), 210:137-147.

Smith, K.L. Jr. and J.M. Teal. 1973. Temperature and pressure effects on respiration of thecosomatous pteropods. Deep-Sea Research, 20:853-858.

van der Spoel, S. 1967. Euthecosomata. J. Noorduijn en Zoon N.V., Gorinchem, The Netherlands.

_____. 1968. The shell and its shape in Cavioliniidae (Pteropoda, Gastropoda). Beaufortia, 15:185-189.

_____. 1972. Notes on the identification and speciation of Heteropoda (Gastropoda). Zoölogische Mededelingen, Leiden, 47:545-560.

_____. 1976. Pseudothecosomata, Gymnosomata and Heteropoda (Gastropoda). Bohn, Scheltema and Holkema, Utrecht, The Netherlands, 484 pp.

Tesch, J.J. 1949. Heteropoda. "Dana" Report, 34:1-53.

_____ 1950. The gymnosomata II. "Dana" Report, 36:1-15.

Thiriot-Quiévreux, C. 1973. Heteropoda. Oceanography and Marine Biology Annual Review, 11:237-261.

Thompson, T.E. and I. Bennett. 1969. *Physalia* nematocysts: utilized by mollusks for defense. Science, 166:1532-1533.

Tunicates

Alldredge, A.L. 1972. Abandoned larvacean houses, a unique food source in the pelagic environment. Science, 177:885-887.

_____. 1976. Field behavior and adaptive strategies of appendicularians (Chordata: Tunicata). Marine Biology, 38:29-39.

_____. 1981. The impact of appendicularian grazing on natural concentrations *in situ*. Limnology and Oceanography, 26:247-257.

Alldredge, A. and L.P. Madin. 1982. Pelagic tunicates: unique herbivores in the marine plankton. Bioscience, 32:655-663.

Baker, A.N. 1971. *Pyrosoma spinosum* Herdman, a giant pelagic tunicate new to New Zealand waters. Records of the Dominion Museum, 7:107-117.

Barham, E.G. 1979. Giant larvacean houses: observations from deep submersibles. Science, 205:1129-1131.

Berner, L.D. 1967. Distributional Atlas of Thaliacea in the California Current Region. CALCOFI Atlas No.8 (California Marine Research Committee), 322 pp.

Blackburn, M. 1979. Thaliacea of the California current region : relations to temperature, chlorophyll, currents, and upwelling. CALCOFI Reports No. 20, pp. 184-214.

Bone, Q. 1997. Biology of Pelagic Tunicates. Oxford University Press, New York.

Bone, Q. and E.R. Trueman. 1983. Jet propulsion in salps (Tunicata, Thaliacea). Journal of Zoology, London, 201:481-506.

_____. 1984. Jet propulsion in *Doliolum* (Tunicata: Thaliacea). Journal of Experimental Marine Biology and Ecology, 76:105-118.

Bone, Q., *et al.* 1980. The communication between individuals in salp chains. I. Morphology of the system. Proceedings of the Royal Society of London, B, 210:549-558.

Cetta, C.M., L.P. Madin and P. Kremer. 1986. Respiration and excretion by oceanic salps. Marine Biology, 91:529-537.

Fenaux, R. 1993. The classification of Appendicularia (Tunicata): history and current state. Mémoires de l'Institut Océanographique. Monaco, 17, vii-123.

Flood, P.R. 1991. Architecture of, and water circulation and flow rate in, the house of the planktonic tunicate *Oikopleura labradoriensis*. Marine Biology, 111:95-111.

Flood, P.R., D. Deibel and C.C. Morris. 1990. Visualization of the transparent, gelatinous house of the pelagic tunicate *Oikopleura vanhoeffeni* using *Sepia* ink. Biological Bulletin, 178:118-125.

Fraser, J.H. 1982. British Pelagic Tunicates. Cambridge University Press, Cambridge, 57 pp.

Galt, C.P., M.S. Grober and P.F. Sykes. 1985. Taxonomic correlates of bioluminescence among appendicularians (Urochordata: Larvacea). Biological Bulletin, 168:125-134.

Hamner, W.M.and B.H. Robison. 1992. *In situ* observations of giant appendicularians in Monterey Bay. Deep-sea Research, Part A, 39:1299-1313.

Harbison, G.R. and R.W. Gilmer. 1976. The feeding rates of the pelagic tunicate *Pegea confederata* and two other salps. Limnology and Oceanography, 21:517-528.

Harbison, G.R. and V.L. McAlister. 1979. The filter-feeding rates and particle retention efficiencies of three species of *Cyclosalpa* (Tunicata: Thaliacea). Limnology and Oceanography, 24:875-892

Harbison, G.R. and R.B. Campenot. 1979. Effects of temperature on the swimming of salps (Tunicata, Thaliacea): implications for vertical migration. Limnology and Oceanography, 24:1081-1091.

Heron, A.C. and E.E. Benham. 1985. Life history parameters as indicators of growth rate in three salp populations. Journal of Plankton Research, 7:365-379.

Hubbard, L.T. Jr. and W.G. Pearcy. 1971. Geographic distribution and relative abundance of Salpidae off the Oregon coast. Journal of the Fisheries Research Board of Canada, 28:1831-1836.

Kremer, P. and L.P. Madin. 1992. Particle retention efficiency of salps. Journal of Plankton Research, 4:1009-1015.

Mackie, G.O. and Q. Bone. 1978. Luminescence and associated effector activity in *Pyrosoma* (Tunicata: Pyrosomida). Proceedings of the Royal Society of London (Series B), 202:483-495.

Madin, L.P. 1974. Field observations on the feeding behavior of salps (Tunicata: Thaliacea). Marine Biology, 25:143-147.

_____. 1990. Aspects of the jet propulsion in salps. Canadian Journal of Zoology, 68:765-777.

Madin, L.P. and G.R. Harbison. 1977. The association of Hyperiidea with gelatinous zooplankton - I. Associations with Salpidae. Deep-Sea Research, 24:449-463.

_____. 1978. Salps of the genus *Pegea* Savigny 1816 (Tunicata: Thaliacea). Bulletin of Marine Science, 28:335-344.

Metcalf, M. M. 1918. The Salpidae: a taxonomic study. United States National Museum, Bulletin, 100, 2(Part 2): 3-193.

Pomeroy, L.R. and D. Deibel. 1980. Aggregation of organic matter by pelagic tunicates. Limnology and Oceanography, 25:643-652.

Purcell, J.E. and L.P. Madin. 1991. Diel patterns of migration, feeding, and spawning by salps in the subarctic Pacific. Marine Ecology Progress Series, 73:211-217.

Ritter, W.E. 1905. The pelagic Tunicata of the San Diego region, excepting the Larvacea. University of California Publications in Zoology, 2:51-112.

Silver, M.W. 1975. The habitat of *Salpa fusiformis* in the California Current as defined by indicator assemblages. Limnology and Oceanography, 20:230-237.

van Soest, R.W.M. 1974. Taxonomy of the subfamily Cyclosalpinae Yount, 1954 (Tunicata, Thaliacea), with descriptions of two new species. Beaufortia, 22:17-55.

_____. 1974. A revision of the genera *Salpa* Forskal, 1775, *Pegea* Savigny, 1816, and *Ritteriella* Metcalf, 1919 (Tunicata, Thaliacea). Beaufortia, 22:153-191.

_____. 1975. Zoogeography and speciation in the Salpidae (Tunicata: Thaliacea). Beaufortia, 23: 181-215.

_____. 1981. A monograph of the order Pyrosomatida (Tunicata, Thaliacea). Journal of Plankton Research, 3:603-631.

Tokioka, T. 1960. Studies on the distribution of appendicularians and some thaliaceans of the north Pacific, with some morphological notes. Publications of the Seto Marine Biology Laboratory, 8:351-443.

Tokioka, T. and L. Berner. 1958. On certain Thaliacea (Tunicata) from the Pacific Ocean, with descriptions of two new species of doliolids. Pacific Science, 12, 317-326.

Yount, J.L. 1954. The taxonomy of the Salpidae (Tunicata) of the Central Pacific Ocean. Pacific Science, 8:276-330.

_____. 1958. Distribution and ecologic aspects of Central Pacific Salpidae (Tunicata). Pacific Science, 12:111-130.

General

Alvariño, A. 1967. Bathymetric distribution of Chaetognatha, Siphonophorae, Medusae and Ctenophorae off San Diego, California. Pacific Science, 21:474-485.

Biggs, D.C. 1977. Respiration and ammonium excretion by open ocean gelatinous zooplankton. Limnology and Oceanography, 22:108-118.

Cairns, S.D., D.R. Calder, A. Brinckmann-Voss, C.B. Castro, P.R. Pugh, C.E. Cutress, W.C. Japp, D.G. Fautin, R.J. Larson, G.R. Harbison, M.N. Arai and D.M Opresko. 1991. Common and scientific names of aquartic invertebrates from the United States and Canada: Cnidaria and Ctenophora. American Fisheries Society Special Publication, 22:1-75, 4 pls.

Hamner, W.M. 1974. Blue-water plankton. National Geographic, 46:530-545.

_____. 1990. Design developments in the planktonkreisel, a plankton aquarium for ships at sea. Journal of Plankton Research, 12:397-402.

Hamner, W.M., L.P. Madin, A.L. Alldredge, R.W. Gilmer and P.P. Hamner. 1975. Underwater observations of gelatinous zooplankton: sampling problems, feeding biology, and behavior. Limnology and Oceanography, 20:907-917.

Harbison, G.R. and L.P. Madin. 1979. Diving - a new view of plankton biology. Oceanus, 22:18-27.

Heine, J.N. 1986. Bluewater diving guidelines. California Sea Grant College Program Publication No. T-CSGCP-014.

Herring, P.J. 1987. Systematic distribution of bioluminescence in living organisms. Journal of Bioluminescence and Chemiluminescence, 1:47-163.

Hyman, L.H. 1940. The Invertebrates: Protozoa through Ctenophora. McGraw-Hill Book Company, New York, 726 pp.

Larson, R.J. 1986. Seasonal changes in standing stocks, growth rates, and production rates of gelatinous predators in Saanich Inlet, BC. Marine Ecology Progress Series, 33:89-98.

_____. 1987. Daily ration and predation by medusae and ctenophores in Saanich Inlet, Canada. Netherlands Journal of Sea Research, 21:35-44.

Laval, P. 1980. Hyperiid amphipods as crustacean parasitoids associated with gelatinous zooplankton. Oceanography and Marine Biology Annual Review, 18:11-51.

Mackie, G.O. 1985. Midwater macroplankton of British Columbia studied by submersible PISCES IV. Journal of Plankton Research, 7:753-777.

Mills, C.E. 1981. Seasonal occurrence of planktonic medusae and ctenophores in the San Juan Archipelago (NE Pacific). Wasmann Journal of Biology, 39:6-29.

_____. 1995. Medusae, siphonophores, and ctenophores as planktivorous predators in changing global ecosystems. ICES Journal of Marine Science, 52:575-581.

Mills, C.E. and J.T. Carlton. 1998. Havens on the high seas: rationale for a system of international reserves for the open ocean. Conservation Biology, in press.

Pearse, V., J. Pearse, M. Buchsbaum, and R. Buchsbaum. 1987. Living Invertebrates. The Boxwood Press, Pacific Grove, CA.

Reid, J.L., G.I. Roden and J.G. Wyllie. 1958. Studies of the California Current system. Contributions from the Scripps Institution of Oceanography, New Series, No. 998, pp. 293-323.

Williamson, J., P. Fenner and J. Burnett. (eds.) 1996. Venomous and poisonous marine animals: a medical and biological handbook. NSW University Press, Sidney Australia, 504 pp.

Etymologies - Derivations of the Scientific Names

Aegina citrea: Named for *Aegina*, youngest daughter of the river god Asopus, who was abducted by Zeus. *Kitrea* is Greek for citron, recalling the deep yellow color of some specimens.

Aequorea spp.: From the Latin *aequoreus*, of a smooth even surface, such as the calm sea.

Aglantha digitale: Combines the Greek *aglan*, glorious, with *anthe*, blossom or bloom. Species name is from the Greek *digitalis*, of the finger, referring to the fingerlike gonads.

Aglaura hemistoma: Derives from the Greek *aglaos*, splendid, shining. The species name combines the Greek *hemi*, half, and *stoma*, mouth.

Aglauropsis aeora: From the Greek *aglaos*, shining, bright, beautiful, and *opsis*, likeness. *Aeora* is from Greek, referring to both the oscillating gonads and the pulsating swimming motion.

Amphinema platyhedos: Derives from the Greek *amphi*, on both sides, and *nema*, thread-members of this genus all have two opposite tentacles. The Greek *platys*, broad, and *hedos*, base, are combined in reference to the broad-based stomach.

Annatiara affinis: The genus name is a rearrangement of another genus name *Tiaranna*, in which this species was originally placed, but later removed. When this species was first described, its systematic position was somewhat unclear, and its Latin species name *affinis*, meaning related to, was applied with the most reasonable choice of genus (later changed, upon seeing more material).

Athorybia rosacea: The genus name may derive from the Greek *athroos*, crowded together. The Latin *rosaceus* means rosette-like, or shaped like a rose.

Atlanta spp.: *Atlanta* (usually *Atalanta*) was a Greek mythological maiden-huntress and Argonaut, said to be the fastest mortal of her time.

Atolla vanhoeffeni: *Atolla* means an island surrounded by coral reefs, apparently in reference to the central lenticular apex surrounded by numerous lappets and tentacular pedalia, but this is not completely clear from the original description. Species named for Ernst Vanhöffen, German biologist of the early 20th century, who studied deep sea medusae in general, and first discovered this species.

Atolla wyvillei: This species was named in honor of Scottish naturalist, Sir. C. Wyville Thomson, scientific director of the 'Challenger' Expedition (1872-1876), on which the first specimens were collected.

Aurelia aurita: Aurelia derives from the old Italian *aurelio*, originally used for what is now called a chrysalis. The species name is from the Latin auritus, *eared*, referring to the shape of the gonads.

Aurelia labiata: The species name is from the Latin *labiatus*, lipped.

Bathochordaeus charon. From the Greek *bathys*, deep, and *chorda*, string of an instrument. *Charon* is the ferryman of Greek mythology who carries souls of the dead across the River Styx.

Bathocyroe fosteri: Combines the Greek *bathys*, deep, with *Ocyroë*, a now-retired genus name for another lobe-flapping ctenophore. The species was named in honor of one of the submersible *Alvin's* pilots, Dudley Foster.

Bathyctena chuni. Combines the Greek *bathys*, deep, and *ktenos*, comb. The species is named for Dr. Carl Chun, late-19th century ctenophore biologist from Leipzig who discovered this species as the first deep-sea ctenophore.

Benthocodon pedunculata. From the Greek *benthos*, depth of the sea, and *kodon*, bell. The species name is from the Latin *pedunculus*, and was given to distinguish this from other species of *Crossota* by its gastric peduncle. The peduncle is now the basis for separating the genus *Benthocodon* from *Crossota*.

Beroe abyssicola. *Beroë* was the mythical daughter of Aphrodite and Butes the Argonaut. The species name is from the Greek *abyssos*, the deep sea, and Latin *-cola*, a suffix meaning inhabitant.

Beroe cucumis. The species name is Latin, meaning cucumber-shaped.

Beroe forskalii. This species is named for Peter Forskål, 18th century Swedish traveler and naturalist who first observed it in the Mediterranean.

Beroe gracilis. *Gracilis* means slender in Latin, describing the distinctive elongation of the body.

Beroe mitrata. The species name derives from the Greek *mitra*, headdress or turban, and recalls the miter-shaped body.

Blackfordia virginica. The genus was named in honor of Eugene G. Blackford, Fish Commissioner of the State of New York, for his contributions to the propagation of food-fishes and to the New York Aquarium, at the end of the 19th century. The species name recalls the type locality in Virginia.

Bolinopsis infundibulum. The genus name means "similar to *Bolina*," once a genus of lobate ctenophore, but now occupied by a mollusc. *Infundibulum* is Latin for funnel.

Bougainvillia spp. The genus was named in honor of Admiral Louis Antoine de Bougainville, French explorer who circumnavigated the world from 1766 to 1769.

Bythotiara stilbosa. Derives from the Greek *bythos*, the depths of the sea, and the Greek *tiara*, turban. *Stilbose* is Greek for glistening, in reference to the exumbrellar nematocysts.

Calycopsis simulans. Combines the Greek *kalyx*, cup, and -opsis, likeness. The species name derives from the Latin *simulo*, imitate or copy, and *-ans*, state or condition of, and reflects the uncertainty of the describer about establishing a separate species for this medusa that is much like *Sibogita geometrica*.

Carinaria cristata. *Carina* is Latin for keel. *Cristata* is Latin for tufted.

Carybdea marsupialis. Named for the Greek *Charybdis*, a whirlpool near Messina. Genus name was originally for the medusa now called *Periphylla periphylla*, in reference to the sometimes sharply-pointed stomach, which was likened to a small inverted whirlpool. The species name derives from the Latin *marsupium*, pouch or purse, and *-alis*, having the nature of, describing the shape of the bell.

Catablema nodulosa. From the Greek, *kata*, down or against, and *blema*, coverlet. The species name is from the Latin, *nodulus*, a small swelling.

Cavolinia tridentata. From the Latin *cavus*, hollow, and *linea*, line. The species name is from the Latin *tri*, three, and *dentis*, tooth, referring to the three projections on the shell.

Cestum veneris. From the Greek *kestos*, girdle. *Veneris* means pertaining to Venus, in Latin.

Charistephane fugiens. Combines the Greek *charis*, loveliness, and *stephanos*, crown. *Fugiens* seems to derive from the Latin *fugio*, to flee, and may reflect the uncertainty when this species was described about whether it was really the larva of another ctenophore.

Chrysaora fuscescens. In Greek mythology, *Chrysaor* was the son of Poseidon and Medusa, his name translating as golden falchion. The species name derives from the Latin *fuscus*, dark or dusky, refering to the dusky bell color.

Chrysaora melanaster. The species name is from the Greek *melas*, black, and *aster*, star, referring to the diagnostic dark pigmentation pattern on a light background.

Cladonema californicum. From the Greek *klados*, for branch, and *nema*, thread, referring to the distinctive branched tentacles. The species is named for its type locality.

Clio pyramidata. Named for *Clio*, a sea nymph and daughter of Oceanus. *Pyramis* is Greek for pyramidal, referring to the shield-shaped shell.

Clione limacina. Genus name seems to be a variation of *Clio*, sea-nymph daughter of Oceanus. Species name is from the Latin *limacis*, snail or slug, and *-ina*, a diminutive suffix.

Cliopsis krohni. Name combines *Clio*, a sea nymph and daughter of Oceanus, with *-opsis*, appearance or likeness, for a variation on the related genus *Clio*. The species is named in honor of the mid-19th century German biologist August Krohn, who worked on pelagic molluscs, as well as many other kinds of plankton.

Clytia gregaria. *Clytia* was a nymph of Greek mythology who loved Helios, the sun. In Latin *gregarius* means pertaining to a flock or herd, and was chosen because this species often occurs in very large numbers at the surface.

Colobonema sericeum. Derives from the Greek *kolobos*, clipped or shortened, and *nema*, thread, presumably referring to the short tentacle stubs characteristic of net-collected specimens, whose tentacles fall off from rough handling. *Sericeus* is Greek for silken.

Corolla calceola. *Corolla* is Latin for small wreath or crown. *Calceolus* is Latin for a small shoe or half boot, which roughly describes the shape of the pseudoconch.

Creseis virgula. *Creseis* (=*Cressida*) is a legendary woman of the medieval Latin period. *Virgula* is Latin for small twig or branch.

Crossota rufobrunnea. From the Greek *krossotos*, fringed. The species name combines Latin *rufus*, red, and *brunneus*, brown, in describing the color of the subumbrella.

Crucibranchaea macrochira. Combines the Latin *crucis*, cross, and Greek *branchos*, gill. The species name is from the Greek *makros*, long, and *cheir*, hand.

Cunina spp. Named for *Cunina*, a Roman Goddess who protected children in the cradle.

Cyanea capillata. From the Greek *kyanos*, dark blue, because the first species described in this genus (*Cyanea lamarcki*) is blue. The species name is from the Latin *capillus*, hair, presumably in reference to its many tentacles.

Cyclosalpa affinis. From the Greek *kyklos*, circle, and refers to the characteristic whorled form of the chain. The species name is from the Latin *affinis*, neighboring, in reference to the aggregate form.

Cyclosalpa bakeri. This species was named in honor of Dr. Fred Baker, San Diego physician, naturalist-conchologist and civic leader, who first invited W. Ritter to move his biological studies to San Diego, establishing that region's prominence in marine science.

Deepstaria enigmatica. Named for the submersible '*Deepstar 4000*' from which the first specimen was collected. The species name refers to the fact that this first specimen, collected in the San Diego Trough, seemed so strange that its precise systematic position could not be determined.

Deiopea kaloktenota. *Deiopea* was a nymph of Greek mythology. The species name is derived from the Greek *kalos*, beautiful, and *ktenion*, comb.

Dendronotus iris. From the Greek *dendron*, tree, and *notos*, back, in reference to the branched cirrata on the back of this animal. *Iris* is Latin for rainbow, recalling the colorful pigmentation of this species.

Desmopterus papilio. Combines the Greek *desmos*, bond or fetter, and *pteron*, wing. *Papilio* is Latin for butterfly.

Dipleurosoma typicum. Combines the Greek *dis*, double, Greek *pleura*, side, and *soma*, body, perhaps referring to its habit of reproduction by binary fission. The species name is from the Latin *typus*, model or example, meaning that this was the type species for the genus.

Dolioletta gegenbauri. The genus name is from Latin, meaning having the nature of a small barrel. The species is named for C. Gegenbaur, German naturalist who studied many planktonic gelatinous organisms in the Mediterranean in the mid-19th century.

Dromalia alexandri. The genus name derives from the Greek *dromos*, a race, race course, running, flight, or swift, and *halos*, the sea.

Dryodora glandiformis. From *dryad*, a wood-nymph in Greek and Latin mythology, combined with the suffix *-orum*, having the nature of. The species name is derived from the Latin *glans*, acorn-shaped, and *forma*, shape.

Eperetmus typus. May derive from the Greek *epeir*, to pull to, in reference to the centripetal canals. The species name is from the Latin *typus*, example, indicating that this is the model for the genus.

Euphysa spp. The genus name seems to be a variation of the Greek *euphues*, shapely or graceful.

Euplokamis dunlapae. From the Greek *eu*, good, and *plokamos*, curl or lock of hair, referring to the unusual tightly-curled side branches of the tentacles. The species is named for Helen L. Dunlap, developmental biologist who studied it in the 1960s.

Eutonina indicans. Derives from the Greek *eutoneo*, energetic, active, elastic, and *-inus*, having the quality of. *Indicans*, refers to the way the peduncle and manubrium are able to point towards food captured by the marginal tentacles.

Fiona pinnata. *Pinnata* is Latin for feathered.

Firoloida desmaresti. The species name combines the Latin *de*, from, and *maris*, sea, but may be a patronym.

Foersteria purpurea. Named in honor of R. Earle Foerster of the University of British Columbia, who studied hydromedusae of the Pacific Coast in the early 20th century. The species name reflects the distinctive purple coloration of living specimens.

Forskalia sp. Named for Peter Forskål, Swedish naturalist and traveler of the 18th century.

Gastropteron pacificum. Derives from the Greek *gastros*, stomach, and *pteron*, wing, recalling its ability to swim. Species name refers to the type locality in the Pacific Ocean (Aleutians).

Geryonia proboscidalis. Named for the three-headed *Geryon* of Greek mythology, grandson of Medusa by Chrysaor. *Proboskis*, is Greek for the trunk of an elephant, recalling its long, trunk-like peduncle.

Glaucus atlanticus. *Glaucus* was a sea-god of Greek mythology. *Atlanticus* recalls the original type locality of this cosmopolitan species.

Gleba cordata. *Gleba* is Latin for a clod or clump. *Chordatus* is Latin for heart-shaped.

Gonionemus vertens. From the Greek, *gonio*, angle or corner, and *nema*, thread, in reference to the tentacles, which are usually bent at the point of the adhesive disk. The species name derives from the Latin *verto*, turn, referring to its peculiar swimming pattern, whereby the medusa swims vigorously to the surface interface, is flipped over, and sinks back down.

Haeckelia beehleri. Named in honor of Ernst H. Haeckel, late-19th century German biologist and natural historian. The species is named for Commodore William H. Beehler, U.S.N. at Key West, who was highly supportive of A. G. Mayer's Tortugas Biological Laboratory.

Haeckelia bimaculata. The species name is from the Latin *bi*, two, and *macula*, spot or stain, referring to the two symmetrical pigmented spots near the tentacle bases.

Haeckelia rubra. The species name means red in Latin, referring to its four characteristic reddish spots.

Haliclystus octoradiatus. The genus name is from the Greek *halos*, sea, and *kleistos*, closed. *Octoradiatus* reflects its 8-part symmetry.

Halicreas minimum. Combines the Greek *halos*, sea, and *kreas*, flesh. *Minimus* is Latin for least or smallest.

Halimedusa typus. Combines the Greek *halos*, sea or salt, and *Medusa*, the mythical Greek Gorgon with serpent locks whose gaze turned men to stone. The species name is from the Latin *typus*, example, indicating that this is the model for the genus.

Haliscera spp. The genus name is from the Greek *halos*, sea, and *keras*, horn or bow.

Halitholus spp. Combines the Greek *halos*, sea, and *tholos*, dome or cupola, referring to the characteristic gelatinous apical projection on the bell.

Halitrephes maasi. Combines the Greek *halos*, sea, and *trepho*, to thicken or to feed or nurture. The species is named in honor of Dr. Otto Maas, Professor of Zoology and expert on medusae at the University of Munich at the turn of the 19th to 20th centuries.

Helicosalpa virgula. Genus name is from the Greek *helikos*, spiral. *Virgula* is Latin for small twig or branch.

Hormiphora californensis. From the Greek *hormos*, necklace, and *phoreus*, bearer, perhaps referring to the studded appearance of the tentacles. The species name recalls the type locality.

Hormiphora cucumis. The species name derives from the Latin *cucumis*, cucumber, but more pertinent is that this species was confused with *Beroe cucumis* and thus now carries the same species name.

Hormiphora palmata. The species name is from the Greek *palmatus*, palm-like, referring to the complex, hand-like tentacle appendages on some species of *Hormiphora*. Unfortunately, this name was applied both to a juvenile specimen with hand-like appendages, and to an adult that has only normal filiform tentacle side branches and the species name ended up staying with the adult, a species which never has the distinctive palmate appendages.

Iasis zonaria. Named for *Iasis* of Greek mythology, father of Atalanta. The Greek *zonaria* means of a belt or girdle, presumably in reference to the chain of aggregates.

Janthina janthina. Genus and species names are from the Greek *ianthinos*, violet-blue.

***Kiyohimea* spp**. From *Kiyohimé*, the heroine of a Japanese legend.

Lampea pancerina. Genus name is a corruption of *Lampetia*, one of the Greek nereids, whose name derives from the Greek *lampe*, lamp. The species name was intentionally combined with this genus to honor the work of the 19th century Italian scientist Paolo Panceri, who studied luminescence in marine animals.

***Leuckartiara* spp**. This genus (originally *Tiara*, but that was found to be preoccupied by a mollusc) was renamed in honor of Rudolf Leuckart, 19th century German zoologist, who among other accomplishments separated the echinoderms from the coelenterates.

Leucothea pulchra. Named for *Leucothea*, white sea-goddess of Rhodes, who rescued Odysseus from drowning. The species name is from the Latin *pulchrum*, beauty.

Limacina helicina. From the Latin *limacis*, snail or slug, and *-ina*, a diminutive suffix. The species name is from the Greek *helix*, whorl or spiral, also combined with *-ina*.

Linuche unguiculata. Combines the Latin *linea*, line, and *uchus*, carry or hold. The species name is from Latin *unguicula*, small claw or talon, probably in reference to its small curved tentacles.

Liriope tetraphylla. The genus name is from the Greek *leirion*, lily. Species combines the Greek *tettares*, four, and *phyllon*, leaf, recalling the four leaf-shaped gonads.

Maeotias inexspectata. Named for the Maeotae, a Scythian tribe living around the Sea of Azov (called "Palus Maeotis" by the Romans). The species name is from the Latin *inexspectatus*, unexpected.

Manania distincta. The species name is from the Latin *distinctus*, separate or different.

Manania handi. This species is named in honor of west coast coelenterate biologist, Professor Cadet Hand, former Director of the Bodega Marine Laboratory.

Manania gwilliami. This species is named in honor of Professor G. Frank Gwilliam of Reed College, who studied this species in detail in the 1950s, as a student of Cadet Hand.

Melibe leonina. *Meliboëa* was a daughter of Niobe in Greek mythology. The species name is from the Greek *leon*, lion, and *-ina*, a diminutive suffix.

Melicertum octocostatum. Seems to derive from the Latin *mellis*, honey, and *certum*, a certainty, apparently in reference to the honey-colored gonads. The species name is from the Latin *octo*, eight, and *costatus*, ribbed, referring to the 8 radial canals and gonads.

Mitrocoma cellularia. Derives from the Greek *mitra*, turban, and *kome*, hair of the head. The species name is in recognition of the very large epithelial cells on the exumbrella.

Mitrocomella polydiademata. The genus name is a diminutive of the related genus *Mitrocoma*. The species name combines the Greek *polys*, many, and *diadem*, crown.

Modeeria rotunda. Named in honor of Adolph Modeer, Swedish 18th century naturalist who made the first attempt toward systematizing the knowledge about medusae. *Rotundus* is Latin for round or spherical, in reference to the thick jelly and general aspect of this species.

Muggiaea atlantica. The species name recalls the type locality.

Nanomia bijuga. May derive from the Greek *nannos*, dwarfish or small and *miar*, stained with blood, referring to the red-stained pneumatophore. The species name is from the Latin *bi*, two, and either *jugum*, yoke, pair, ridge, in reference to the shape of the nectophores, or *jugo*, to join, referring to the biserial arrangement of the nectophores.

Nausithoe atlantica. From the Greek *naus*, ship, and *thoë*, penalty. *Nausithous* was a son of Odysseus. The species name recalls the type locality in the North Atlantic.

Neoturris breviconis. *Turris* is Latin for tower–the genus *Neoturris* was erected to replace the older genus name *Turris*, which was already occupied by a mollusc. The species name combines the Latin *brevis*, short, and Greek *konos*, cone, referring to the short, broad apical projection.

Obelia **spp**. The genus name derives from the Greek *obelias*, a kind of bread or cake cooked on a spit.

Ocyropsis maculata. A corruption of the earlier genus name *Ocyroe*, which was preoccupied and no longer available for a ctenophore, and derives from the Greek *okys*, swift and *opsis*, appearance. *Macula* is Latin for spot.

Oikopleura **spp**. From the Greek *oikos*, house, and *pleura*, side, referring to the house surrounding the animal.

Pantachogon **spp**. The genus name appears to be some kind of adjective formation from the Greek *pantachou*, everywhere.

Paraphyllina **spp**. From the Greek *para*, beside or near, and *phyllon*, leaf; *-ina* is a diminutive suffix.

Pegantha **spp**. The genus name combines the Greek *pege*, a spring, and *anthe*, a flower.

Pegea confoederata. From the Greek *pege*, water. The species name combines the Latin *con*, together, and *foederis*, association, referring to the aggregate stage.

Pelagia colorata. From the Greek *pelagios*, of the sea. *Coloratus* is Latin for colored or variegated, acknowledging its vividly pigmented bell.

Pelagia noctiluca. *Pelagios* is Greek for of the sea. *Noctiluca* means "the one shining by night" in Latin, in reference to its bioluminescence.

Periphylla periphylla. Derives from the Greek *peri*, around or near, and *phyllon*, leaf.

Phacellophora camtschatica. Derives from the Greek *phakelos*, bundle or cluster, and *phoreus*, bearer or carrier. This species is named in recognition of its type location off the Kamchatka Peninsula, in the Russian Far-East.

Phylliroe atlantica. From the Greek *phyllon*, leaf, and *rhoë*, stream or flowing. *Atlantica* recalls the locality where this species was originally collected.

Phyllorhiza punctata. From the Greek *phyllon*, leaf, and *rhiza*, root. The species name is from the Latin *punctum*, spot or dot, recalling its spotted bell.

Physalia physalis. Genus and species names are from the Greek *physallis*, a bladder or bubble.

Physophora hydrostatica. Combines the Greek *physa*, bellows or bubble, and *phoresis*, bearing. The species name is from the Greek *hydor*, water, and *statikos*, standing or resting.

Pleurobrachia bachei. From the Greek *pleura*, side, and *brachion*, arm. The species is named for Professor A. D. Bache, superintendent of the Coast Survey, who gave Alexander Agassiz a summer job as an engineer on a ship surveying the United States/Canada boundary in the Strait of Georgia, where this species was first collected.

Pneumodermopsis **sp**. Combines the Greek *pneuma*, air or breath, *dermatos*, skin, and *-opsis*, appearance or likeness.

Polyorchis haplus. Combines the Greek *polys*, many, and *orchis*, testicle. The species name is from the Greek *haploos*, simple, in recognition that the radial canals in most of the specimens are unbranched.

Polyorchis penicillatus. The species name is a dimutive of the Latin *penis*, meaning either tail or a male copulatory organ, and *-atus*, having the nature of, probably in reference to the pendant gonads.

Porpita porpita. From the Greek *porpe*, buckle, pin or brooch, likening this floating sea creature to a piece of jewelry.

Praya dubia. From the name of the principal town in the Cape Verde Islands, Praia, where the first specimens were collected. *Dubius* is Latin for doubtful.

Proboscidactyla flavicirrata. From the Greek *proboskis*, snout or trunk, and *daktylos*, finger, apparently referring to the manubrium and its highly folded lips. The species name is from the Latin *flavidus*, yellowish, and *cirratus*, fringed or curly, in reference to the tentacles.

Pterotrachea **spp**. Combines the Greek *pteron*, wing, and *tracheia*, windpipe.

Ptychogastria polaris. From the Greek *ptychos*, fold or leaf, and *gastros*, stomach, in recognition of its 8-partite stomach. *Polaris* recalls the type location in the Arctic.

Ptychogena **spp**. The genus name is derived from the Greek *ptychos*, fold or leaf, and *genos*, race or stock, in reference to its folded gonads.

Pyrosoma atlanticum. From the Greek *pyros*, fire and *soma*, body, acknowledging the brilliant bioluminescence produced by these animals. *Atlanticum* recalls the type locality of this species, which is now known to have worldwide distribution.

Rathkea octopunctata. One of the first examples of this species was collected by H. Rathke in the Black Sea-the genus was later named after him. The species name is from the Latin *octo*, eight, and *punctata*, spotted, referring to the 8 darkly pigmented tentacle bulbs.

Rhizophysa **sp**. From the Greek *rhiza*, root, and *physa*, bellows or bubble.

Salpa fusiformis. *Salpe* was the Greek name for a kind of marine fish. The species name is from the Latin *fusus*, spindle and *forma*, shape, recalling the distinctive fusiform shape of the aggregate individuals.

Salpa maxima. The Latin *maximus*, greatest, refers to the potentially large size of this species.

***Sarsia* spp**. This genus was named in honor of Michael Sars, Norwegian priest and professor of zoology in the mid-19th century, who originally described S. *tubulosa* (in another genus).

Scrippsia pacifica. The genus was named for members of the Scripps family of San Diego, who had recently endowed a new building in La Jolla for the Marine Biological Association of San Diego (soon to be renamed the Scripps Institution for Biological Research, and later the Scripps Institution of Oceanography). The species name recalls the type locality in La Jolla, California.

***Solmaris* spp**. Combines the Latin *sol*, sun, and *maris*, sea, in reference to its overall sunburst shape.

***Solmissus* spp**. From the Latin *sol*, sun, and *missus*, sent.

Stomolophus meleagris. From Greek *stoma*, mouth, and *lophos*, crest. *Meleagris* is Greek for guinea fowl, or peacock.

Stomotoca atra. From the Greek *stoma*, mouth, and *tokas*, birth. The species name is from the Latin *ater*, black, in reference to its usually dark gonads.

Stygiomedusa gigantea. The genus name means a medusa belonging to the lower world, deriving from the river Styx. The species name refers to its very large size.

***Sulculeolaria* sp**. From the Latin *sulculus*, a small furrow or groove, and *-arius*, having the nature of, possibly combined with *laros*, dainty.

Tetraplatia volitans. The genus name is from the Greek *tessares* (*tettares*), four, and *plateia*, street. The species name derives from the Latin *volitans*, one flying about.

Tetrorchis erythrogaster. From the Greek *tessares* (*tettares*), four, and *orchis*, testicle. The species name combines the Greek *erythros*, red, and *gastros*, stomach.

Thalassocalyce inconstans. Combines the Greek *thalassa*, sea, with *kalyx*, cup. The species name is from the Latin *inconstans*, changeable.

Thalia democratica. *Thalia* was one of the muses (of comedy) of Greek mythology. The species name combines the Greek *demos*, people, and *kratos*, strength, in oblique reference to the many individuals in a chain.

Thetys vagina. *Thetys*, or *Thetis*, was one of the Nereids, a Greek sea nymph who dressed in seaweed. The species name *vagina* is Latin for sheath or scabbard, probably referring to the open shape of a salp.

***Thliptodon* sp**. The genus name combines the Greek *thlipsis*, pressure, and *odon*, tooth.

Tiaropsidium kelseyi. The name *Tiaropsidium* was chosen as a variation of *Tiaropsis* when the latter genus was divided in two. The species name is in honor of Mr. F. W. Kelsey, Secretary of the Marine Biological Association of San Diego in the early 20th century.

Tiaropsis multicirrata. Combines the Greek *tiara*, turban, and the suffix *-opsis*, likeness. The species name is from the Latin *multus*, much, and *cirratus*, fringed or curled, referring to the large number of marginal tentacles.

Vallentinia adherens. Named in honor of Mr. Rupert Vallentin, who first collected this genus as part of a general biological collection in the Falkland Islands, 1898-1899. The species name refers to its habit of clinging to algae with the adhesive tentacles.

Velamen parallelum. From the Latin *velamen*, cover or garment. The species name is from the Latin *parallelus*, side by side, probably referring to the location of the comb rows.

Velella velella. Derives from the Latin *velum*, sail, and *-ellus*, a diminutive suffix.

Weelia cylindrica. Named in honor of Dr. P.B. van Weel, of the University of Hawaii, who assisted J.L. Yount in his study of central Pacific salps. The species name is from the Greek *kylindros*, roller, in reference to the general shape of solitary individuals.

Classification of the Gelatinous Zooplankton of the Pacific Coast of North America

Phylum Cnidaria / Class Hydrozoa

Subclass Hydroidomedusae (Hydromedusae)
Order Anthomedusae
Family Bougainvilliidae
Bougainvillia britannica (Forbes, 1841)
Bougainvillia multitentaculata Foerster, 1923
Bougainvillia principis (Steenstrup, 1850)
Bougainvillia superciliaris (L. Agassiz, 1849)
Chiarella centripetalis Maas, 1897
Family Calycopsidae
Bythotiara stilbosa Mills & Rees, 1979
Calycopsis simulans (Bigelow, 1909)
Family Halimedusidae
Halimedusa typus Bigelow, 1916
Family Pandeidae
Amphinema platyhedos
        Arai & Brinckmann-Voss, 1983
Annatiara affinis (Hartlaub, 1913)
Catablema multicirrata Kishinouye, 1910
Catablema nodulosa (Bigelow, 1913)
Halitholus cirratus Hartlaub, 1913
Halitholus pauper Hartlaub, 1913
Leuckartiara spp.
Neoturris breviconis (Murbach & Shearer, 1902)
Stomotoca atra A. Agassiz, 1862
Family Rathkeidae
Rathkea octopunctata (M. Sars, 1835)
Family Cladonematidae
Cladonema californicum Hyman, 1947
Cladonema radiatum Dujardin, 1843
Family Corynidae
Sarsia spp.
Family Moerisiidae
Maeotias inexspectata Ostroumoff, 1896
Family Polyorchidae
Polyorchis haplus Skogsberg, 1948
Polyorchis penicillatus (Eschscholtz, 1829)
Scrippsia pacifica Torrey, 1909
Family Euphysidae
Euphysa flammea (Linko, 1905)
Euphysa japonica (Maas, 1909)
Euphysa tentaculata Linko, 1905
Family Velellidae
Porpita porpita (Linnaeus, 1758)
Velella velella (Linnaeus, 1758)

Order Leptomedusae
Family Aequoreidae
Aequorea aequorea (Forskål, 1775)
Aequorea flava (A. Agassiz, 1862)
Aequorea forskalea Péron & Lesueur, 1809
Aequorea victoria (Murbach & Shearer, 1902)
Aequorea sp.

Family Blackfordiidae
Blackfordia virginica Mayer, 1910

Family Dipleurosomatidae
Dipleurosoma typicum Boeck, 1866
Family Melicertidae
Melicertum octocostatum (M. Sars, 1835)
Family Eirenidae
Eutonina indicans (Romanes, 1876)
Family Laodiceidae
Ptychogena californica Torrey, 1909
Ptychogena lactea A. Agassiz, 1865
Family Tiarannidae
Modeeria rotunda (Quoy & Gaimard, 1827)
        (= *Tiaranna rotunda*)
Family Mitrocomidae
Foersteria purpurea (Foerster, 1923)
Mitrocoma cellularia (A. Agassiz, 1865)
Mitrocomella polydiademata (Romanes, 1876)
Family Tiaropsidae
Tiaropsidium kelseyi Torrey, 1909
Tiaropsis multicirrata (M. Sars, 1835)
Family Campanulariidae
Clytia gregaria (A. Agassiz, 1862)
        (= *Phialidium gregarium*)
Obelia spp.

Order Limnomedusae
Family Olindiasidae
Aglauropsis aeora Mills, Rees & Hand, 1976
Eperetmus typus Bigelow, 1915
Gonionemus vertens A. Agassiz, 1862
Vallentinia adherens Hyman, 1947
Family Proboscidactylidae
Proboscidactyla flavicirrata Brandt, 1835

Order Narcomedusae
Family Aeginidae
Aegina citrea Eschscholtz, 1829
Family Solmarisidae
Pegantha spp.
Solmaris spp.
Family Cuninidae
Cunina spp.
Solmissus spp.
Solmundella bitentaculata (Quoy & Gaimard, 1833)

Order Trachymedusae
Family Geryoniidae
Geryonia proboscidalis (Forskål, 1775)
Liriope tetraphylla (Chamisso & Eysenhardt, 1821)

Family Halicreatidae
Halicreas minimum Fewkes, 1882
Haliscera bigelowi Kramp, 1947
Haliscera conica Vanhöffen, 1902
Halitrephes maasi Bigelow, 1909
Halitrephes valdiviae Vanhöffen, 1912
Family Ptychogastriidae
Ptychogastria polaris Allman, 1878
Family Rhopalonematidae
Aglantha digitale (O.F. Müller, 1776)
Aglaura hemistoma Péron & Lesueur, 1809
Benthocodon pedunculata (Bigelow, 1913)
Colobonema sericeum Vanhöffen, 1902
Colobonema typicum (Maas, 1897)
Crossota alba Bigelow, 1913
Crossota rufobrunnea (Kramp, 1913)
Pantachogon spp.
Tetrorchis erythrogaster
Vampyrocrossota childressi Thuesen, 1993

Subclass Siphonophorae
Order Cystonecta
Family Physaliidae
Physalia physalis (Linnaeus, 1758)
Family Rhizophysidae
Rhizophysa sp.

Order Physonecta
Family Apolemiidae
Apolemia uvaria (Lesueur, ?1811)
Family Agalmidae
Nanomia bijuga (Chiaje, 1841)
Family Physophoridae
Physophora hydrostatica Forskål, 1775
Family Athorybiidae
Athorybia rosacea (Forskål, 1775)
Family Rhodaliidae
Dromalia alexandri Bigelow, 1911
Family Forskaliidae
Forskalia edwardsi Kölliker, 1853

Order Calycophora
Family Prayidae
Praya sp.
Family Diphyidae
Muggiaea atlantica Cunningham, 1892
Sulculeolaria sp.

Class Cubozoa/Cubomedusae
Family Carybdeidae
Carybdea marsupialis (Linnaeus, 1758)

Class Scyphozoa / Scyphomedusae

Order Stauromedusae
Suborder Eleutherocarpida
Family Lucernariidae
Haliclystus octoradiatus (Lamarck, 1816)
Haliclystus salpinx Clark, 1863
Haliclystus sanjuanensis Hyman, 1940
Haliclystus stejnegeri Kishinouye, 1899

Suborder Cleistocarpida
Family Depastridae
Manania distincta (Kishinouye, 1910)
Manania gwilliami Larson & Fautin, 1989
Manania handi Larson & Fautin, 1989

Order Coronatae
Family Atollidae
Atolla vanhoeffeni Russell, 1957
Atolla wyvillei Haeckel, 1880
Family Linuchidae
Linuche unguiculata (Swartz, 1788)
Family Nausithoidae
Nausithoe ?atlantica Broch, 1914
Nausithoe punctata Kölliker, 1853
Family Paraphyllinidae
Paraphyllina intermedia Maas, 1903
Paraphyllina ransoni Russell, 1956
Paraphyllina rubra Neppi, 1915
Family Periphyllidae
Periphylla periphylla (Péron & Lesueur, 1809)
Family Tetraplatidae
Tetraplatia volitans Busch, 1851

Order Semaeostomeae
Family Pelagiidae
Chrysaora fuscescens Brandt, 1835
Chrysaora melanaster Brandt, 1835
Pelagia colorata Russell, 1964
Pelagia noctiluca (Forskål, 1775)
Family Cyaneidae
Cyanea capillata (Linnaeus, 1758)
Family Ulmaridae
Subfamily Aureliinae
Aurelia aurita (Linnaeus, 1758)
Aurelia labiata Chamisso & Eysenhardt, 1821
Aurelia limbata Brandt, 1835
Subfamily Sthenoniinae
Phacellophora camtschatica Brandt, 1835
Subfamily Poraliinae
Poralia sp.
Subfamily Stygiomedusinae
Stygiomedusa gigantea (Browne, 1910)
Subfamily Deepstariinae
Deepstaria enigmatica Russell, 1967

Order Rhizostomeae
Family Mastigiidae
Phyllorhiza punctata von Lendenfeld, 1884
Family Stomolophidae
Stomolophus meleagris L. Agassiz, 1862

Phylum Ctenophora
Order Cydippida
Family Haeckeliidae
Haeckelia beehleri (Mayer, 1912)
Haeckelia bimaculata C. Carré & D. Carré, 1989
Haeckelia rubra (Kölliker, 1853)

Family Bathyctenidae
Bathyctena chuni (Moser, 1909)
Family Lampeidae
Lampea pancerina (Chun, 1879)
Family Pleurobrachiidae
Hormiphora californensis (Torrey, 1904)
Hormiphora undescribed species
Hormiphora cucumis (Mertens, 1833)
Hormiphora palmata Chun, 1898
Pleurobrachia bachei A. Agassiz, 1860
Family Euplokamidae
Euplokamis dunlapae Mills, 1987
Family Mertensiidae
Charistephane fugiens Chun, 1879
Undescribed species of mertensiid ctenophore
Family Dryodoridae
Dryodora glandiformis (Mertens, 1833)

Order Thalassocalycida
Family Thalassocalycidae
Thalassocalyce inconstans Madin & Harbison, 1978

Order Lobata
Family Bathocyroidae
Bathocyroe ?fosteri Madin & Harbison 1978
Family Bolinopsidae
Bolinopsis infundibulum (O. F. Müller, 1776)
Family Deiopeidae
Deiopea kaloktenota Chun, 1879
Kiyohimea spp.
Family Leucotheidae
Leucothea pulchra Matsumoto, 1988
Family Ocyropsidae
Ocyropsis maculata (Rang, 1828)

Order Cestida
Family Cestidae
Cestum veneris Lesueur, 1813
Velamen parallelum (Fol, 1869)

Order Beroida
Family Beroidae
Beroe abyssicola Mortensen, 1927
Beroe ?cucumis Fabricius, 1780
Beroe forskalii Milne Edwards, 1841
Beroe gracilis Künne, 1939
Beroe mitrata (Moser, 1907)

Phylum Mollusca
Class Gastropoda
Subclass Prosobranchia
Order Mesogastropoda / Suborder Ptenoglossa
Family Janthinidae
Janthina janthina (Linnaeus, 1758)

Superfamily Heteropoda
Family Atlantidae
Atlanta spp.
Family Carinariidae
Carinaria cristata (Linnaeus, 1766)
Family Pterotracheidae
Firoloida desmaresti Lesueur, 1817
Pterotrachea coronata Forsskål, 1775
Pterotrachea scutata Gegenbaur, 1855

Subclass Opisthobranchia
Order Cephalaspidea
Family Gastropteridae
Gastropteron pacificum Bergh, 1894

Order Thecosomata
Suborder Euthecosomata
Family Limacinidae
Limacina helicina (Phipps, 1774)
Family Cavoliniidae
Cavolinia inflexa (Lesueur, 1813)
Cavolinia tridentata (Niebuhr, 1775)
Clio pyramidata Linnaeus, 1767
Creseis acicula (Rang, 1828)
Creseis virgula (Rang, 1828)
Suborder Pseudothecosomata
Family Peraclididae
Peraclis spp.
Family Cymbuliidae
Corolla calceola (Verrill, 1880)
Gleba cordata Niebuhr, 1776
Family Desmopteridae
Desmopterus papilio Chun, 1889

Order Gymnosomata / Suborder Gymnosomata
Family Pneumodermatidae
Crucibranchaea macrochira (Meisenheimer, 1905)
Pneumodermopsis sp.
Family Cliopsidae
Cliopsis krohni Troschel, 1854
Family Clionidae
Clione limacina (Phipps, 1774)
Thliptodon sp.

Order Nudibranchia
Suborder Dendronotacea
Family Dendronotidae
Dendronotus iris Cooper, 1863
Family Phylliroidae
Phylliroe atlantica Bergh, 1871
Family Tethyidae
Melibe leonina (Gould, 1852)
Suborder Aeolidacea
Family Glaucidae
Glaucus atlanticus Forster, 1777

Family Fionidae
Fiona pinnata (Eschscholtz, 1831)
Phylum Chordata
Subphylum Urochordata/Tunicata
Class Thaliacea
Order Salpida
Family Salpidae
Cyclosalpa affinis (Chamisso, 1819)
Cyclosalpa bakeri Ritter, 1905
Helicosalpa virgula (Vogt, 1854)
Iasis zonaria (Pallas, 1774)
Pegea confoederata (Forskål, 1775)
Pegea socia (Bosc, 1802)
Salpa fusiformis Cuvier, 1804
Salpa maxima Forskål, 1775
Thalia democratica (Forskål, 1775)
Thetys vagina Tilesius, 1802
Weelia cylindrica Cuvier, 1804 (= *Salpa cylindrica*)

Order Doliolida
Family Doliolidae
Dolioletta gegenbauri Uljanin, 1884

Order Pyrosomatida
Family Pyrosomatidae
Pyrosoma atlanticum (Péron, 1804)

Class Larvacea /Appendicularia
Order Copelata
Family Oikopleuridae
Bathochordaeus charon Chun, 1900
Oikopleura dioica Fol, 1872
Oikopleura labradoriensis Lohmann, 1892
Oikopleura longicauda (Vogt, 1854)
Oikopleura vanhoeffeni Lohmann, 1896
Family Fritillariidae
Fritillaria borealis Lohmann, 1896

Habitats of Gelatinous Zooplankton of the Pacific Coast of North America

(A question mark before the genus name indicates that the habitat is not certain. A question mark before the species name indicates that the species identification is not certain.)

COASTAL SPECIES

HYDROMEDUSAE
Aglantha digitale
Aglauropsis aeora
Aequorea aequorea var. *albida*
Aequorea forskalea
Aequorea macrodactyla
Aequorea victoria
Aequorea sp.
Bougainvillia bougainvillei
Bougainvillia britannica
Bougainvillia multitentaculata
Bougainvillia muscus (=Bougainvillia ramosa)
Bougainvillia nordgaardi
Bougainvillia principis
Bougainvillia superciliaris
Bythotiara huntsmani
Bythotiara stilbosa
Catablema multicirrata
Catablema nodulosa
Chiarella centripetalis
Clytia gregaria
Clytia lomae
Dipleurosoma typicum
Dipurena bicircella
Eirene mollis
Eperetmus typus
Euphysa flammea
Euphysa japonica
Euphysa tentaculata
Euphysora bigelowi
Eutonina indicans
Geomackiea zephyrolata
Halimedusa typus
Halitholus pauper
Halitholus sp. I
Halitholus sp. II
Hybocodon prolifer
Hydrocoryne bodegensis
Laodicea sp.
Leuckartiara foersteri
Leuckartiara nobilis
Leuckartiara octona
Leuckartiara sp.
Lizzia ferrarii
Melicertum octocostatum
Mitrocoma cellularia
Mitrocomella polydiademata
Neoturris breviconis

Obelia dichotoma
Obelia geniculata
Orthopyxis compressa
Phialella fragilis
Phialella zappai
Plotocnide borealis
Polyorchis penicillatus
Proboscidactyla circumsabella
Proboscidactyla flavicirrata
Proboscidactyla occidentalis
Proboscidactyla ornata
Ptychogena californica
Ptychogena lactea
Rathkea octopunctata
Sarsia apicula
Sarsia cliffordi
Sarsia coccometra
Sarsia eximia
Sarsia japonica
Sarsia princeps
Sarsia tubulosa
Sarsia viridis
Sarsia sp. A
Scrippsia pacifica
Staurophora mertensii
Stomotoca atra
Stomotoca pterophylla
Tiaropsidium kelseyi
Tiaropsis multicirrata
Trichydra pudica
Vallentinia adherens
Zanclea costata

SIPHONOPHORES
Chelophyes appendiculata
Chelophyes contorta
Dimophyes arctica
Muggiaea atlantica
Muggiaea bargmannae
Nanomia bijuga
Sphaeronectes gracilis

CUBOMEDUSAE
Carybdea marsupialis

SCYPHOMEDUSAE
Aurelia aurita
Aurelia labiata
Aurelia limbata

Chrysaora achlyos
Chrysaora fuscescens
Chrysaora melanaster
Craterolophus sp.
Cyanea capillata
Haliclystus octoradiatus
Haliclystus salpinx
Haliclystus sanjuanensis
Haliclystus stejnegeri
Kyopoda lamberti
Linuche unguiculata
Lychnorhiza sp.
Manania distincta
Manania gwilliami
Manania handi
Manania hexaradiata
Nausithoe punctata
Phacellophora camtschatica
Phyllorhiza punctata
Stomolophus meleagris

CTENOPHORES
Bolinopsis ?infundibulum
Bolinopsis ?vitrea
Pleurobrachia bachei

MOLLUSCS
Dendronotus spp.
Gastropteron pacificum
Melibe leonina

TUNICATES
Doliolum nationalis
Fritillaria borealis
Megalocercus huxleyi
Oikopleura dioica
Oikopleura labradoriensis
Oikopleura longicauda
Oikopleura vanhoeffeni
Stegosoma magnum

SPECIES CHARACTERISTIC OF PROTECTED BAYS AND INLETS

HYDROMEDUSAE
Blackfordia virginica
Catablema multicirrata
Cladonema californicum

Cladonema myersi
Cladonema° radiatum
Cladonema uchidai
Gonionemus vertens
Halimedusa typus
Maeotias inexpectata
Polyorchis haplus
Polyorchis penicillatus

SCYPHOMEDUSAE
Aurelia aurita
Aurelia labiata
Manania handi

VERY LOW SALINITY SPECIES

HYDROMEDUSAE
Blackfordia virginica
Craspedacusta sowerbyi
Maeotias inexpectata
Moerisia sp.

SLOPE/OCEANIC EPIPELAGIC SPECIES

HYDROMEDUSAE
Aegina citrea
Aeginopsis laurentii
Aequorea aequorea var. *albida*
Aglantha digitale
Aglaura hemistoma
? Aglauropsis aeora
Amphinema ?platyhedos
Calycopsis nematophora
Calycopsis simulans
Climacocodon ikarii
Clytia gregaria
Cunina duplicata
Cunina frugifera
Cunina octonaria
Cunina peregrina
Cunina spp.
? Eperetmus typus
Eutonina indicans
Geryonia proboscidalis
Heterotiara anonyma
Liriope tetraphylla
Neoturris pileata
Pegantha martagon
Phialopsis diegensis
Porpita porpita
Rhopalonema velatum
Sarsia princeps
? Scrippsia pacifica

Sibogita geometrica
Solmaris ?leucostyla
Solmissus marshalli
Solmundella bitentaculata
Staurophora mertensii
Velella velella

SIPHONOPHORES
Abyla bicarinata
Abyla haeckeli
Abylopsis eschscholtzi
Abylopsis tetragona
Agalma elegans
Agalma okeni
Amphicaryon ernesti
Apolemia uvaria
Athorybia rosacea
Bassia bassensis
Ceratocymba dentata
Chelophyes appendiculata
Chelophyes contorta
Diphyes bojani
Diphyes dispar
Eudoxoides mitra
Eudoxoides spiralis
Forskalia edwardsi
Halistemma rubrum
Hippopodius hippopus
Lensia campanella
Lensia challengeri
Lensia conoidea
Lensia cossack
Lensia fowleri
Lensia hotspur
Lensia multicristata
Lensia subtilis
Lensia subtiloides
Lilyopsis rosea
Muggiaea atlantica
Nanomia bijuga
Physalia physalis
Physophora hydrostatica
Praya dubia
Praya reticulata
Rhizophysa eysenhardti
Rosacea cymbiformis
Sphaeronectes gracilis
Stephanophyes superba
Sulculeolaria biloba
Sulculeolaria chuni
Sulculeolaria monoica
Sulculeolaria quadrivalvis
Sulculeolaria turgida
Vogtia glabra
Vogtia spinosa

SCYPHOMEDUSAE
Chrysaora fuscescens
Chrysaora melanaster
? Chrysaora achlyos
Cyanea capillata
Linuche unguiculata
Nausithoe punctata
Pelagia colorata
Pelagia noctiluca
Tetraplatia volitans

CTENOPHORES
Beroe cucumis
Beroe ?cucumis
Beroe forskalii
Beroe gracilis
Beroe mitrata
Cestum veneris
Deiopea kaloktenota
Dryodora glandiformis
Eurhamphaea vexilligera
Haeckelia beehleri
Haeckelia bimaculata
Haeckelia rubra
Hormiphora californensis
Hormiphora cucumis
Hormiphora palmata
Lampea pancerina
Leucothea pulchra
Ocyropsis maculata
Velamen parallelum
Thalassocalyce ?inconstans

MOLLUSCS
Atlanta californiensis
Atlanta gaudichaudi
Atlanta helicinoides
Atlanta inclinata
Atlanta inflata
Atlanta lesueuri
Atlanta peroni
Atlanta turriculata
Cardiopoda placenta
Cardiopoda richardi
Carinaria cristata forma *japonica*
Carinaria galea
Carinaria lamarcki
Cavolinia globosa
Cavolinia inflexa
Cavolinia tridentata
Cavolinia uncinata
Clio cuspidata
Clio pyramidata
Clio recurva
Cliopsis krohni

Corolla calceola=? Corolla spectabilis
Creseis acicula
Creseis virgula
Crucibranchaea macrochira
Cuvierina columnella
Diacria quadridentata
Diacria trispinosa
Desmopterus pacificus
Desmopterus papilio
Fiona pinnata
Firoloida desmaresti
Glaucus atlanticus
Gleba cordata
Hyalocylis striata
Janthina janthina
Janthina prolongata
Limacina bulimoides
Limacina helicina
Limacina inflata
Limacina lesueuri
Limacina trochiformis
Notobranchaea macdonaldi
Phylliroe atlantica
Pneumoderma sp.
Pneumodermopsis ciliata
Provotella sp.
Pterotrachea coronata
Pterotrachea hippocampus
Pterotrachea minuta
Pterotrachea scutata
Spongiobranchia australis
Styliola subula
Thliptodon sp.

TUNICATES

Brooksia rostrata
Cyclosalpa affinis
Cyclosalpa bakeri
Cyclosalpa pinnata
Cyclosalpa strongylentron
Dolioletta gegenbauri
Dolioletta gegenbauri var. *tritonis*
Doliolina mulleri
Doliolina obscura
Doliolina rarum
Doliolina undulatum
Doliolum denticulatum
Doliolum nationalis
Doliopsis rubescens
Helicosalpa virgula
Iasis zonaria
Ihlea punctata
Kowalevskia tenuis

Oikopleura albicans
Oikopleura cornutogastra
Oikopleura fusiformis
Oikopleura longicauda
Oikopleura rufescens
Pegea confoederata
Pegea socia
Pyrosoma atlanticum
Ritteriella picteti
Salpa fusiformis
Salpa maxima
Thalia democratica
Thetys vagina
Traustedtia multitentaculata
Weelia cylindrica

DEEP WATER SPECIES

HYDROMEDUSAE

Aegina citrea
Aegina sp. A
Aeginopsis laurentii
Aeginura grimaldii
Aglantha digitale
Amphinema platyhedos
Amphogona apicata
Annatiara affinis
Benthocodon hyalinus
Benthocodon pedunculata
Botrynema brucei
Bythotiara depressa
Colobonema sericeum
Colobonema typicum
Crossota alba
Crossota rufobrunnea
Crossota sp. A
Cunina frugifera
Euphysa sp.
Euphysora gigantea
Euphysora valdiviae
Foersteria purpurea
Geomackiea zephyrolata
Halicreas minimum
Haliscera bigelowi
Haliscera conica
Haliscera racovitzae
Halitrephes maasi
Halitrephes valdiviae
Heterotiara anonyma
Leuckartiara sp.
Merga reesi
Modeeria rotunda
Neoturris fontata

Pandea rubra
Pantachogon ?haeckeli
Pantachogon sp. A (orange)
Pantachogon sp. B (? *scotti*)
Pegantha sp.
Ptychogastria polaris
Ptychogena lactea
Rhopalonema funerarium
Solmaris quadrata
Solmissus incisa
Solmissus marshalli
Solmundella bitentaculata
Sminthea eurygaster
Tetrorchis erythrogaster
Tetrorchis sp. A
Tima sp.
Vampyrocrossota childressi
Zanclea costata

SIPHONOPHORES

Agalma elegans
Amphicaryon acaule
Amphicaryon ernesti
Bargmannia elongata
Bathyphysa sp.
Cordagalma cordiforme
Chuniphyes moserae
Chuniphyes multidentata
Clausophyes galeata
Clausophyes moserae
Crystallophyes amygdalina
Desmophyes annectens
Dimophyes arctica
Dromalia alexandri
Enneagonum hyalinum
Erenna richardi
Forskalia edwardsi
Frillagalma vityazi
Gilia reticulata
Halistemma amphytridis
Halistemma rubrum
Heteropyramis maculata
Lensia achilles
Lensia ajax
Lensia baryi
Lensia challengeri
Lensia conoidea
Lensia exeter
Lensia grimaldi
Lensia havock
Lensia hostile
Lensia lelouveteau
Lensia meteori
Lensia multicristata

Maresearsia praeclara
Marrus orthocanna
Nanomia bijuga
Nectadamas diomedeae
Nectopyramis natans
Nectopyramis thetis
Physophora hydrostatica
Praya dubia
Praya reticulata
Ramosia vitiazi
Rosacea plicata
Vogtia glabra
Vogtia kuruae
Vogtia pentacantha
Vogtia serrata
Vogtia spinosa

SCYPHOMEDUSAE
Atolla parva
Atolla vanhoeffeni
Atolla wyvillei
Atorella vanhoeffeni
Deepstaria enigmatica
Lucernaria sp.
Nausithoe atlantica
Nausithoe ?rubra
Paraphyllina intermedia
Paraphyllina ransoni
Paraphyllina rubra
Periphylla periphylla
Periphyllopsis galatheae
Poralia rufescens
Stygiomedusa gigantea
Tetraplatia volitans

CTENOPHORES
Bathocyroe ?fosteri
Bathyctena chuni
Beroe abyssicola
Beroe ?cucumis
Bolinopsis infundibulum

Charistephane fugiens
Euplokamis dunlapae
Euplokamis sp.
Hormiphora californensis
Hormiphora undescribed species
Kiyohimea aurita
Kiyohimea usagi
Lampea sp.
Thalassocalyce ?inconstans
undescribed species of mertensiid

MOLLUSCS
Carinaria cristata forma japonica
Carinaria galea
Carinaria lamarcki
Clio polita
Clio pyramidata
Cliopsis krohni
Cymbulia peroni
Gleba cordata
Peraclis apicifulva
Peraclis bispinosa
Peraclis reticulata
Thliptodon sp.

TUNICATES
Bathochordaeus charon
Cyclosalpa affinis
Cyclosalpa bakeri
Iasis zonaria
Pyrosoma atlanticum
Salpa fusiformis

Index to Common and Scientific Names